Plant Cell Biology

Plant Cell Biology

Nadia Fallon

SYRAWOOD
PUBLISHING HOUSE
New York

Published by Syrawood Publishing House,
750 Third Avenue, 9th Floor,
New York, NY 10017, USA
www.syrawoodpublishinghouse.com

Plant Cell Biology
Nadia Fallon

© 2022 Syrawood Publishing House

International Standard Book Number: 978-1-64740-067-5 (Hardback)

Cataloging-in-Publication Data

Plant cell biology / Nadia Fallon.
    p. cm.
Includes bibliographical references and index.
ISBN 978-1-64740-067-5
1. Plants--Cytology. 2. Cytology. 3. Plant cells and tissues. I. Fallon, Nadia.
QK725 .P53 2022
581.072 4--dc23

Table of Contents

Preface

This book has been written, keeping in view that students want more practical information. Thus, my aim has been to make it as comprehensive as possible for the readers. I would like to extend my thanks to my family and co-workers for their knowledge, support and encouragement all along.

Plant cells are the eukaryotic cells having true nucleus along with the specialized structures known as organelles. A few of these organelles are chloroplast, cell wall, and ribosomes. Chloroplasts are the special organelles within plant cells that create sugars through the process of photosynthesis. Plant cells have cell walls that give the cell strength and maintain high turgidity. They are composed of cellulose, pectin and hemicelluloses. Some of the common types of plant cell are parenchyma and collenchyma. The living cells that perform various functions such as storage, support to photosynthesis and phloem loading are known as parenchyma. Collenchyma cells have thickened cellulosic cell walls and are alive at maturity. This book aims to shed light on some of the unexplored aspects of plant cell biology. It is an upcoming field of science that has undergone rapid development over the past few decades. Those in search of information to further their knowledge will be greatly assisted by this book.

A brief description of the chapters is provided below for further understanding:

Chapter – Introduction

Plant cell is the eukaryotic cell that is present in green plants. Plant cell biology deals with the study of structure, diagram, types and functions of plant cells. This chapter has been carefully written to provide an easy understanding of plant cell biology.

Chapter – Plant Tissue and its Types

Plant tissue is a group of cells which are similar in structure, origin and functions. It includes merismatic tissue, vascular tissue, ground tissue, dermal tissue, secretory plant tissue, etc. The topics elaborated in this chapter will help in gaining a better perspective about these types of plant tissues.

Chapter – Plant Cell Organelles

A plant cell is enclosed by a cell wall, containing a membrane-bound nucleus and other cell organelles. Protoplasm, cytoplasm, plastids, tannosome, etc. are some of these organelles. This chapter closely examines these plant cell organelles to provide an extensive understanding of the subject.

Chapter – The Plant Hormones

Plant hormones are molecules produced within the plants in low concentrations. A few of these hormones are auxin, gibberlin, cytokinin, jasmonate, karrikin, florigen, ethylene, peptide, strigolactone, etc. All these different plant hormones have been carefully analyzed in this chapter.

Chapter – Photosynthesis

Photosynthesis is the process by which plants prepare their food in the presence of sunlight, carbon dioxide and chlorophyll. It occurs as two types of reactions – light dependent reactions and light independent reactions. This chapter discusses about these reactions and related aspects of photosynthesis in detail.

Nadia Fallon

1
Introduction

Plant cell is the eukaryotic cell that is present in green plants. Plant cell biology deals with the study of structure, diagram, types and functions of plant cells. This chapter has been carefully written to provide an easy understanding of plant cell biology.

Plant Cell

Plant cells are the basic unit of life in organisms of the kingdom Plantae. They are eukaryotic cells, which have a true nucleus along with specialized structures called organelles that carry out different functions. Plant cells have special organelles called chloroplasts which create sugars via photosynthesis.

Animals, fungi, and protists also have eukaryotic cells, while bacteria and archaea have simpler prokaryotic cells. Plant cells are differentiated from the cells of other organisms by their cell walls, chloroplasts, and central vacuole. The chloroplasts within plant cells can undergo photosynthesis, to produce glucose. In doing so, the cells use carbon dioxide and they release oxygen.

Other organisms, such as animals, rely on this oxygen and glucose to survive. Plants are considered autotrophic because they produce their own food and do not have to consume any other organisms. Specifically, plant cells are photoautotrophic because they use light energy from the sun to produce glucose. Organisms that eat plants and other animals are considered heterotrophic.

The other components of a plant cell, the cell wall and central vacuole, work together to give the cell rigidity. The plant cell will store water in the central vacuole, which expands the vacuole into the sides of the cell. The cell wall then pushes against the walls of other cells, creating a force known as turgor pressure. Turgor pressure between cells allows plants to grow tall and reach more sunlight.

Plant Cell Structure

Like the fungi, another kingdom of eukaryotes, plant cells have retained the protective cell wall structure of their prokaryotic ancestors. The basic plant cell shares a similar construction motif with the typical eukaryote cell, but does not have centrioles, lysosomes, intermediate filaments, cilia, or flagella, as does the animal cell. Plant cells do, however, have a number of other specialized structures, including a rigid cell wall, central vacuole, plasmodesmata, and chloroplasts. Although plants (and their typical cells) are non-motile, some species produce gametes that do exhibit flagella and are, therefore, able to move about.

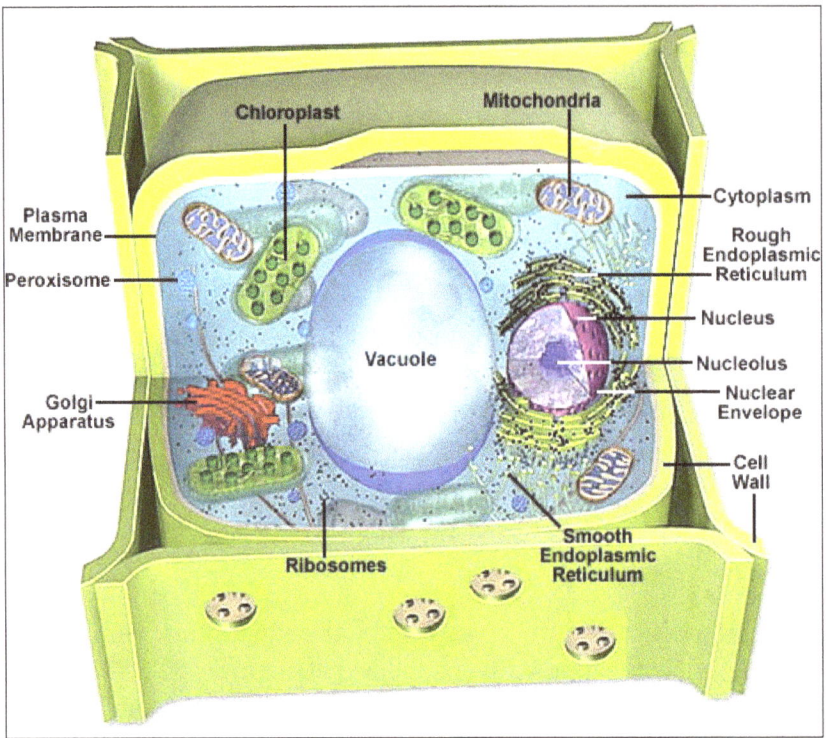

Plants can be broadly categorized into two basic types: vascular and nonvascular. Vascular plants are considered to be more advanced than nonvascular plants because they have evolved specialized tissues, namely xylem, which is involved in structural support and water conduction, and phloem, which functions in food conduction. Consequently, they also possess roots, stems, and leaves, representing a higher form of organization that is characteristically absent in plants lacking vascular tissues. The nonvascular plants, members of the division Bryophyta, are usually no more than an inch or two in height because they do not have adequate support, which is provided by vascular tissues to other plants, to grow bigger. They also are more dependent on the environment that surrounds them to maintain appropriate amounts of moisture and, therefore, tend to inhabit damp, shady areas.

It is estimated that there are at least 260,000 species of plants in the world today. They range in size and complexity from small, nonvascular mosses to giant sequoia trees, the largest living organisms, growing as tall as 330 feet (100 meters). Only a tiny percentage of those species are directly used by people for food, shelter, fiber, and medicine. Nonetheless, plants are the basis for the Earth's ecosystem and food web, and without them complex animal life forms (such as humans) could never have evolved. Indeed, all living organisms are dependent either directly or indirectly on the energy produced by photosynthesis, and the byproduct of this process, oxygen, is essential to animals. Plants also reduce the amount of carbon dioxide present in the atmosphere, hinder soil erosion, and influence water levels and quality.

Plants exhibit life cycles that involve alternating generations of diploid forms, which contain paired chromosome sets in their cell nuclei, and haploid forms, which only possess a single set. Generally these two forms of a plant are very dissimilar in appearance. In higher plants, the diploid generation, the members of which are known as sporophytes due to their ability to produce spores, is usually dominant and more recognizable than the haploid gametophyte generation. In Bryophytes, however, the gametophyte form is dominant and physiologically necessary to the sporophyte form.

Animals are required to consume protein in order to obtain nitrogen, but plants are able to utilize inorganic forms of the element and, therefore, do not need an outside source of protein. Plants do, however, usually require significant amounts of water, which is needed for the photosynthetic process, to maintain cell structure and facilitate growth, and as a means of bringing nutrients to plant cells. The amount of nutrients needed by plant species varies significantly, but nine elements are generally considered to be necessary in relatively large amounts. Termed macroelements, these nutrients include calcium, carbon, hydrogen, magnesium, nitrogen, oxygen, phosphorus, potassium, and sulfur. Seven microelements, which are required by plants in smaller quantities, have also been identified: boron, chlorine, copper, iron, manganese, molybdenum, and zinc.

Thought to have evolved from the green algae, plants have been around since the early Paleozoic era, more than 500 million years ago. The earliest fossil evidence of land plants dates to the Ordovician Period (505 to 438 million years ago). By the Carboniferous Period, about 355 million years ago, most of the Earth was covered by forests of primitive vascular plants, such as lycopods (scale trees) and gymnosperms (pine trees, ginkgos). Angiosperms, the flowering plants, didn't develop until the end of the Cretaceous Period, about 65 million years ago—just as the dinosaurs became extinct.

- Cell Wall - Like their prokaryotic ancestors, plant cells have a rigid wall surrounding the plasma membrane. It is a far more complex structure, however, and serves a variety of functions, from protecting the cell to regulating the life cycle of the plant organism.

- Chloroplasts - The most important characteristic of plants is their ability to photosynthesize, in effect, to make their own food by converting light energy into chemical energy. This process is carried out in specialized organelles called chloroplasts.

- Endoplasmic Reticulum - The endoplasmic reticulum is a network of sacs that manufactures, processes, and transports chemical compounds for use inside and outside of the cell. It is connected to the double-layered nuclear envelope, providing a pipeline between the nucleus and the cytoplasm. In plants, the endoplasmic reticulum also connects between cells via the plasmodesmata.

- Golgi Apparatus - The Golgi apparatus is the distribution and shipping department for the cell's chemical products. It modifies proteins and fats built in the endoplasmic reticulum and prepares them for export as outside of the cell.

- Microfilaments - Microfilaments are solid rods made of globular proteins called actin. These filaments are primarily structural in function and are an important component of the cytoskeleton.

- Microtubules - These straight, hollow cylinders are found throughout the cytoplasm of all eukaryotic cells (prokaryotes don't have them) and carry out a variety of functions, ranging from transport to structural support.

- Mitochondria - Mitochondria are oblong shaped organelles found in the cytoplasm of all eukaryotic cells. In plant cells, they break down carbohydrate and sugar molecules to provide energy, particularly when light isn't available for the chloroplasts to produce energy.

- Nucleus - The nucleus is a highly specialized organelle that serves as the information processing and administrative center of the cell. This organelle has two major functions: it stores the cell's hereditary material, or DNA, and it coordinates the cell's activities, which include growth, intermediary metabolism, protein synthesis, and reproduction (cell division).

- Peroxisomes - Microbodies are a diverse group of organelles that are found in the cytoplasm, roughly spherical and bound by a single membrane. There are several types of microbodies but peroxisomes are the most common.

- Plasmodesmata - Plasmodesmata are small tubes that connect plant cells to each other, providing living bridges between cells.

- Plasma Membrane - All living cells have a plasma membrane that encloses their contents. In prokaryotes and plants, the membrane is the inner layer of protection surrounded by a rigid cell wall. These membranes also regulate the passage of molecules in and out of the cells.

- Ribosomes - All living cells contain ribosomes, tiny organelles composed of approximately 60 percent RNA and 40 percent protein. In eukaryotes, ribosomes are made of four strands of RNA. In prokaryotes, they consist of three strands of RNA.

- Vacuole - Each plant cell has a large, single vacuole that stores compounds, helps in plant growth, and plays an important structural role for the plant.

Leaf Tissue Organization - The plant body is divided into several organs: roots, stems, and leaves. The leaves are the primary photosynthetic organs of plants, serving as key sites where energy from light is converted into chemical energy. Similar to the other organs of a plant, a leaf is comprised of three basic tissue systems, including the dermal, vascular, and ground tissue systems. These three motifs are continuous throughout an entire plant, but their properties vary significantly based upon the organ type in which they are located.

Plant Cell Biology

Plant cell biology is the study of all types of plant cells. Plant cells are the basic building block of plant life, and they carry out all of the essential functions necessary for plant survival. Plant cell biology is particularly concerned with structure, growth, signalling, differentiations and death of plant cells.

2
Plant Tissue and its Types

Plant tissue is a group of cells which are similar in structure, origin and functions. It includes merismatic tissue, vascular tissue, ground tissue, dermal tissue, secretory plant tissue, etc. The topics elaborated in this chapter will help in gaining a better perspective about these types of plant tissues.

Plant tissue is a collection of similar cells performing an organized function for the plant. Each plant tissue is specialized for a unique purpose, and can be combined with other tissues to create organs such as leaves, flowers, stems and roots. The following is a brief outline of plant tissues, and their functions within the plant.

Permanent Tissues in Plants

Simple Tissues

Simple tissues are composed of same type of cells, and thus are homogeneous in nature. The complex tissues, on the other hand, are heterogeneous, being composed of different types of cells. The third group of permanent tissue is the (c) secretory or special tissue. In higher plants, tissues are very seldom homogeneous from morphological or physiological point of view.

There are three types of simple permanent tissues in plants. They are:

- Parenchyma,
- Collenchyma,
- Sclerenchyma.

Parenchyma

The term parenchyma refers to tissues which shows little specialisation and concerned with various physiological functions of the plant. The cells have the power of wound recovery and regeneration. A small group of parenchyma cells or even a single cell may be cultured and differentiated into a whole flowering plant in synthetic medium.

Phylogenetically, the tissue is considered to be primitive as the multicellular plants of the lower groups consist only of parenchyma. Ontogenetically, parenchyma is also considered primitive as its cells are morphologically alike to those of meristems.

It is the most common simple tissue of the plants and occupies major portions of the plant body. Plant life begins with the formation of the parenchymatous tissue by the embryo. The tissue is composed usually of isodiametric cells with intercellular spaces.

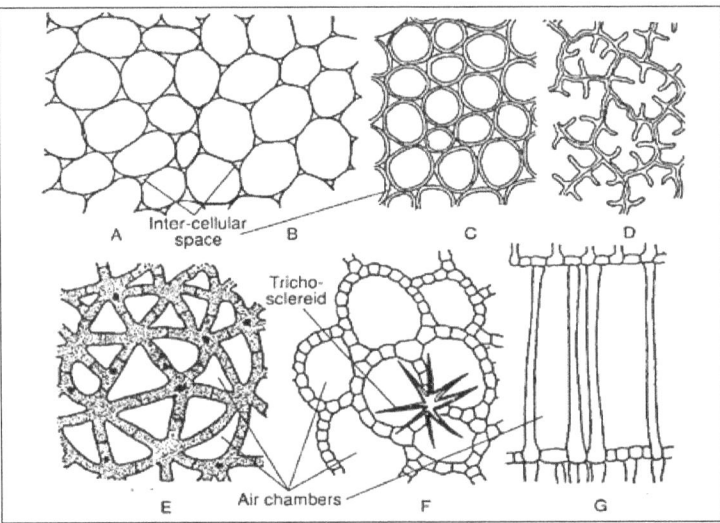

Parenchyma tissues: A. Thin-walled parenchyma in sunflower cortex. B. Same from pith. C. Thick-walled parenchyma in Clematis pith. D. Same from Pinus needle. E. Aerenchyma from Musa petiole. F. Same from Nymphaea petiole. G. Same from Jussiaea stem.

The cells have functional protoplasts. This tissue is ubiquitous in all plants and generally occupies the softer portions of the plant body like epidermis, cortex, pith, mesophyll tissue, pulp of the fleshy fruits, embryo and endosperm of the seeds etc.

Parenchyma tissue is the fundamental tissue of the plant body as it provides the ground for other tissues. Bodies of lower plants are made up of parenchyma cells. The meristems are also parenchymatous in nature. Thus parenchyma tissue is the precursor of all other tissues. So, the parenchyma tissue is considered to be the most primitive tissue, both phylogenetically and ontogenetically.

Collenchyma

Collenchyma is another primary permanent simple tissue consisting of elongated living cells with uneven cellulosic walls. It is the supporting tissue of the plant. Ontogenetically, it develops from certain elongated cells resembling procarnbium which are formed in the very early stages of differentiation of the meristem.

Younger cells of tissue show more extensibility and plasticity than the older ones which are less plastic, harder and more brittle. Collenchyma cells may also contain chloroplasts and carry out photosynthesis. In some cases collenchyma cells store tannins as secondary metabolites. The tissue is actually thick-walled parenchyma specialised to give mechanical support to the plants.

Sclerenchyma

Sclerenchyma is the third type of simple permanent tissue. Cells of this tissue are longer. Cell walls of this tissue are very thick, hard and lignified with very little water due to secondary deposition. The lumen is almost obliterated. At maturity, cells of the sclerenchyma tissue become dead. Simple pits or slightly bordered pits are found over the walls.

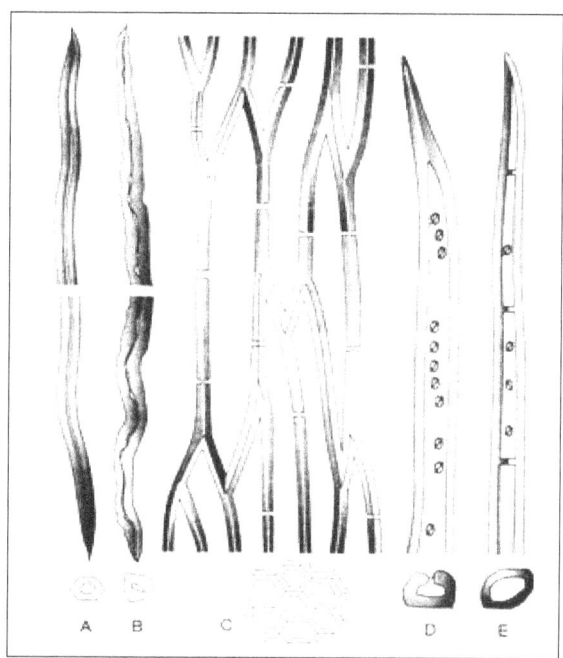

Sclerenchyma fibres in L.S. and T.S.: A & B Fibres with highly thickened walls. C, A bunch of fibres with interlocked ends and simple pits. D. A fibre tracheid. E. A septate fibre tracheid.

Cells of the sclerenchyma tissue differ in shape, structure, origin and development. Different transitional stages are found between the various cell shapes. For that reason, though it is difficult to classify them, but, in general, sclerenchyma is divided into fibres and sclereids.

Fibres are generally elongated cells and sclereids as short cells. The former originate from merisematic cells whereas the latter develop from parenchyma cells due to secondary deposition of the wall materials.

Fibres

These are very long and narrow cells with tapered and sometimes branched ends. Walls are uniformly thickened and highly lignified. The pits are small, round or slit-like in outline. The cell lumen is small due to much thickening of the secondary wall and may be continuous or septate.

The ends may be blunt or branched with interlocked arrangements. The fibres are always dead at maturity. They are angular in outline after cross section. In certain cases the fibre walls are cellulosic and non- lignified. Some fibres have mucilagenous walls.

Distribution

Fibres remain distributed in different organs of the plant body. In the leaflets of Cycas they occur singly as idioblasts. They may occur in separate strands in the cortex or as bundle caps associated with vascular bundles or in the xylem and the phloem as wood and bast fibres.

Fibres are classified into two groups based on their positions in the plant body:

- Xylary fibres, and
- Extraxylary fibres.

Xylary Fibres

These are also known as intraxylary fibres or wood fibres. These are an integral part of the xylem and originate from the same meristematic tissues. Xylary fibres show variations in their shape, size, wall thickness and pitting pattern.

The pits are simple or bordered in nature. These fibres possess lignified secondary walls. There are two main types of xylary fibres – libriform fibres and fibre-tracheids, based on the wall-thickness, type and amount of pits.

Libriform Fibres

The term is derived from the word liber meaning inner bark as they resemble phloem fibres and are usually longer than the tracheids. Walls of these fibres are extremely thick with reduced simple pits. Sometimes in libriform fibres the pit canal becomes elongated and the inner pit aperture becomes slit-like. The inner pit apertures of a pit-pair are usually at right angles to each other.

Fibre-Tracheids

These are intermediate between libriform fibres and tracheids. Cell walls of the fibre-tracheids are of medium thickness with bordered pits. The inner pit opening is slitlike. These structures are regarded as reduced tracheids Like libriform fibres the fibre-tracheids may be septate (septate fibres).

Gelatinous or mucilaginous fibre is observed in the secondary xylem of dicotyledonous plants. The innermost layer of the secondary wall of such fibres is made of β-cellulose called G-layer which after absorbing water almost covers the entire cell lumen. This less compact and porous G-layer irreversibly shrinks on drying.

Some elongated cells with thin secondary walls and living protoplasts occur in the secondary xylem, which may be confused with the living libriform fibres and fibre-tracheids. These elongated cells are called substitute fibres.

Extraxylary Fibres

The fibres present anywhere in the plant body other than xylem tissue are called extraxylary fibres or bast fibres. They may remain distributed in the cortex, pericycle and phloem. These fibres are usually long with tapered, blunt or branched ends. The cell walls of these fibres are thick, lignified or non-lignified with simple or slightly bordered pits.

Extraxylary fibres usually may form isolated strands or continuous bands in the cortex and peri-cycle. They may remain as caps above the vascular bundles. These fibres usually occur as patches in mono- cotyledonous leaves.

There are certain fibres termed septate fibres which are found in the xylem and the phloem even of the same species (e.g. Vitis). These fibres are characterized by the presence of partition walls (sep-ta) and protoplasts with plasmodesmatal connections. The septum consists of a middle lamella and two primary wall-like layers and does not fuse with the fibre wall.

In longitudinal section the fibre tips are viewed as pointed. These fibres may store starch, oil drop-lets, resins and sometimes calcium oxalate. Secondary xylem of many dicotyledons possesses sep-tate fibres. Non-vascular septate fibres are found in some monocotyledons.

Functions

Presence of the fibres in the different parts of the plant body is mainly to give mechanical strength and rigidity to the plant body as well as to withstand strains and stresses.

Ontogenetically, fibres originate from different meristems such as procambium, cambium, ground meristem and protoderm. Cambium derived fibres develop from the elongated fusiform initials and do not elongate further during maturation. But those derived from the short initials elongate greatly at the time of maturation. Of course, the elongation is very gradual and may take a few months.

Evolution

Evolutionary, fibres have developed from tracheids as many transitional forms between these two types of elements are found in some angiosperms. During the course of evolution the tracheid wall has become thickened, the number of pits and the size of the pit chamber have been reduced and ultimately the bordered pits disappeared.

Commercial Fibres

Commercially there are two types of fibres – hard fibres and soft fibres. The hard fibres are stiff and lignified as found in the leaves of Agave, Yucca, Musa textilis, etc. Soft fibres are extraxylary fibres.

They are soft and flexible and may be lignified or non-lignified as found in jute, flax, hemp, ramie etc. The cotton fibres which are also known as surface fibres as they are produced from the testa of seeds represent the most important commercial fibres.

According to their use the fibres may be grouped as:

- Textile fibres (used in the manufacture of cloths; e.g., cotton, flax, ramie and hemp),

- Cordage fibres (used in making different types of cords or ropes; e.g., jute, cotton, hemp, flax, Musa, Agave etc.),

- Brush fibres (used in the manufacture of brushes and brooms; e.g., Agave, fibres from the Palmae and the inflorescences of Sorghum vulgare etc.),

- Filling fibres (used in stuffing furnitures, mattresses, life-belts etc.; e.g., Ceiba pentandra, cotton, jute, Tillandsia usneoides).

Sclereids

Sclereids or sclerotic cells are non-prosenchymatous, isodiametric or irregular in shape. They normally become dead at maturity. They occur as hard masses of cells within soft parenchyma tissue in many different places in the plant body and are much shorter than true fibres in length. They are the major components in the shell of walnuts and seed coats of pea and many other plants.

In many cases, sclereids may be readily distinguished from the surrounding cells by their shape, size and wall thickness which are called idioblasts. In some cases the sclereids are of very peculiar shapes. The sclereid walls generally possess reniform simple pits with branched canals. The hard and thick walls are lignified and also may be cutinised or suberised.

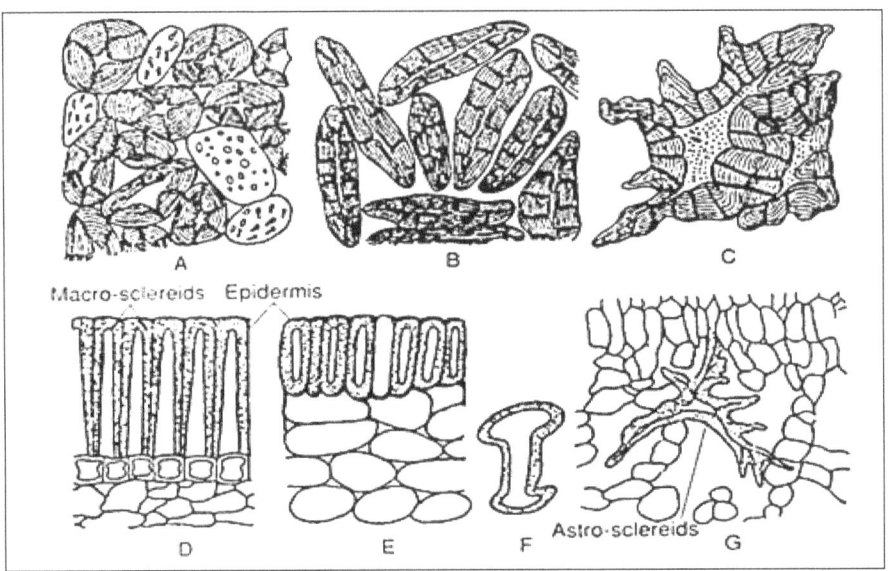

Different type of sclereids: A Brachysclereids from fruit flesh o Pyrus. B. Same from coconut. C. Irregular shaped Sclereids from Tsuga. D. Macrosclereids from the Phaseolus epidemis. E. Same from Alium epidermis. F. Ostecsctereid from Pisum seed coat. G. Asteroscierids from tea leaf.

Distribution

Sclereids are abundantly present in cortex, phloem, pith, mesophyll tissue etc. as idioblasts or as cluster of cells. They are found in fruit pericarp of Pyrus, Psidium etc. They also occur in the seed coats of many plants either singly or in groups.

Types

According to the shape, size and nature of wall thickening the sclereids may be of the following types:

Brachysclereids or Stone Cell

These are more or less isodiametric in appearance. They are also called stone cells or grit cells as they give gritty texture of the fruit flesh of many plants (e.g., Pyrus, Psidium etc.). Brachysclereids are usually found in the phloem, the cortex and the bark of stems.

Macrosclereids

These are rod-shaped columnar sclereids. They often form a continuous palisade like epidermal layer in the testa of seeds in Leguminosae. Macrosclereids also occur in the pulp of Malus sylvestris.

Osteosclereids

These are bone- or spool-shaped sclereids. They are also columnar in arrangement. The ends of these sclereids are enlarged, lobed, or somewhat branched. Such sclereids are mainly found in seed coats (e.g., Pisum) and leaves of certain dicotyledons (e.g., Pisum, Hakea, etc.).

Astrosclereids

These are branched and often star-shaped, mainly found in leaves and stems of many dicotyledonous plants like Thea, Nymphaea, Trchodendron etc.

Trichosclereids

This fifth type of sclereids are very elongated, hair-like, and always single branched sclereids.

Sclereids usually develop from parenchymatous small thin-walled initials by secondary thickening of the cell wall. During brachysclereid development the inner surface of the wall decreases and pits start to develop on the outside of the secondary wall In the early stages of development of the branched sclereid, it begins to branch and ultimately acquires the form of the mature sclereid.

The branches of the sclereid penetrate into the intercellular spaces. The secondary wall becomes thickly deposited in numerous concentric layers with the formation of branched pits. The mode of development of all types of sclereids is common excepting the degree of pitting. The probable cause of such sclerification may be physiological disturbances or ageing.

Table: Difference between sclereids and fibres.

	Sclreids	Fibres
1.	Sclereids are developed from parenchyma cells due to secondary growth in thickness and lignin deposition on their walls.	These are developed from meristematic cells. The cell walls are lignified.
2.	These are mostly isodiametric but may also show various shapes.	Fibres are elongated with tapering ends the usually interlock with each other.
3.	Depending on shapes, they are termed brachy-, macro-, astro-, trichosclereids etc.	There is no major variation in shape which may be septate or asepted.

4.	The sclereids have long pits with rounded pit aperture.	Fibres have pits with slit-like pit aperture.
5.	Ramiform pits are present on brachysclereids.	Ramiform pits are absent on the fibres walls.
6.	Sclereids are found in fruits, barks, pith, cortex, mesophyll tissue, Seed coat etc. and rarely in association of xylem and phloem.	Fibres are present in leaves, stems, roots and fruits, in cortex pith, xylem and phloem, bundle caps, peri-cycle etc.
7.	Sclereids are responsible for mechanical rigidity and elasticity.	Fibres give more mechanical strength to the plant body.

Complex Tissues

Complex tissues are composed of two or more types of cells. Xylem and Phloem are complex tissues. They together comprise the vascular tissue system, the main function of which is conduction of water and minerals from the root xylem to the leaves and prepared food from the leaves to the different parts of the plant body.

Xylem

This complex tissue is composed of both living and non-living cells.

The main components of the tissue are:

- Tracheary elements (vessels and tracheids),
- Xylem parenchyma, and
- Xylary fibres or wood fibres.

Xylem may be primary and secondary depending on the origin. The former one is derived from the procambium whereas the latter is derived from the vascular cambium during secondary growth. The vessel elements of the primary xylem, which differentiate and mature earlier during ontogeny, are known as the protoxylem and those which mature later are called metaxylem.

Tracheary Elements

Tracheary elements are the main conducting elements.

There are two types of tracheary elements:

- Tracheids, and
- Vessels or tracheae.

These elements are non-living. Instead of water conduction they give mechanical support to the plant.

Tracheids

The tracheids are elongated thick-walled cells with tapering ends and remain parallel to the long axis of the organ where they occur. The ends may also be obtuse or rounded, and chisel-like or oblique. It is dead and devoid of protoplast at maturity. The cell lumen is comparatively larger. The hard and lignified cell wall contains bordered pits and annular, spiral, scalari- form and reticulate thickenings.

In cross section, the cell outline appears angular or polyhedral, or sometimes round. The tracheids remain in the tissue one above the other with overlapping ends.

The transverse walls are not obliterated, rather persist with many perforations and the thickening is same as that in the lateral walls. They make communications with surrounding cells through pits. The pits remain on the lateral walls. Bordered pits are present between the adjacent tracheids.

The pit pairs between tracheid and xylem parenchyma may commonly be simple, bordered or half-bordered. In case of half bordered pit the border is formed on the tracheid cell wall and simple pitting on the parenchyma cell wall. Tracheids originate from procambium and vascular cambium.

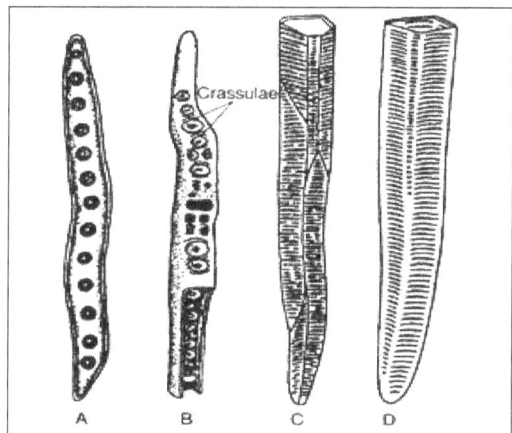

Tracheids: A. Bordered pitted tracheid, B. Bordered pits with crassulae on a portion of a tracheid, C & D. Parts of tracheids with scalariform thickening.

Distribution

Tracheids are found in the primary and secondary xylems of the vascular plants. They predominantly occur in pteridophytes, gymnosperms and primitive angiosperms.

Functions

The functions of tracheid are conduction of water and mechanical support. Very rarely the tracheids store water.

Origin of Tracheid

In primary xylem tracheids originate from procambium and in secondary xylem they originate from cambium ring. A single fusiform initial forms a tracheid.

Phylogeny of Tracheid

Generally, vascular cryptogams have very long tracheids and gymnosperms have tracheids of intermediate length. Different types of secondary wall are found in tracheids the annular, helical, scalariform, reticulate and circular bordered pitted. Ontogenetically the annular elements precede the helical elements. The helical bands again became joined at certain areas forming scalariform elements. Phylogenetically pitted elements are most advanced.

Tracheids with annular thickening have been reported from one of the oldest fossils Cooksonia, a leafless vascular plant from upper Silurian deposits. Baragawanathia another very early leafy vascular plant — also contained the annularly thickened tracheids.

Phylogenetically, vessels have evolved from tracheids and are polyphyletic in origin. Vessels are found in pteridophyta (Selaginella, Equisetum and fern like Actinopteris, Pteridium, Regnellidium and Marsilea) and in gymnosperm (Ephedra, Welwitschia and Gnetum). Apart from angiosperms, the vessels found in the above mentioned genera of pteridophyta and gymnosperm are considered as anomalies.

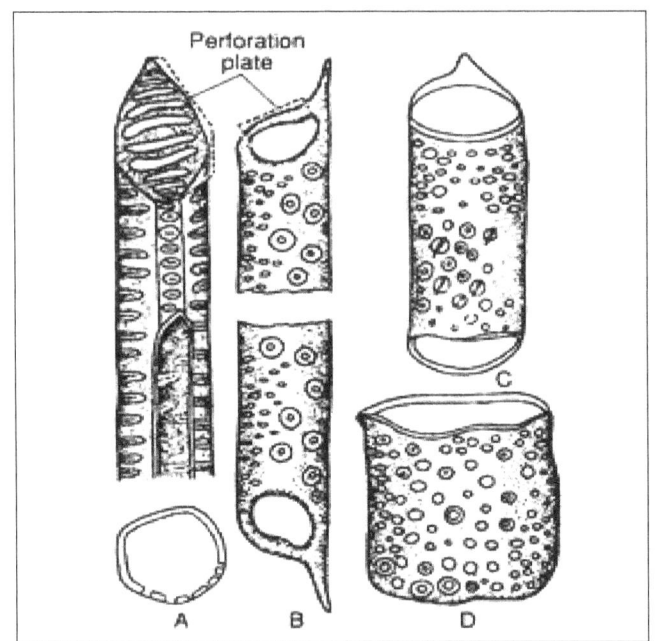

Different types of tracheae: A. With multiple perforation plate in L.S. and T.S. B, C and D. with simple perforation plate.

Vessels

Vessels or tracheae are also thick- walled, lignified non-living members of the xylem tissue. They remain arranged in vertical rows and form a tube-like structure with their perforated end walls. They also have numerous pits on their lateral walls. Water conduction is their main function.

The vessels are dead at maturity. Each vessel is an elongated cylindrical structure. Sometimes their diameter is greater than their length. The perforations on their end walls are large. The vessel elements are parallel with the long axis of the organ in which they occur. The length of the vessel varies and it may be as long as three metres (e.g., Fraxinus).

The perforated end walls of vessel members are called perforation plates, which may also remain on the lateral walls. So their positions are usually terminal, but sometimes sub-terminal or lateral.

The pattern of perforations may be:

- Simple with single large pore at the end (e.g., Quercus);

- Scalariform with several parallel elongated transverse pores (e.g., Liriodendron);

- Foraminate with several more or less circular pores in groups (e.g., Ephedra; and (Reticulate with a network of small pores (e.g., Rhoeo).

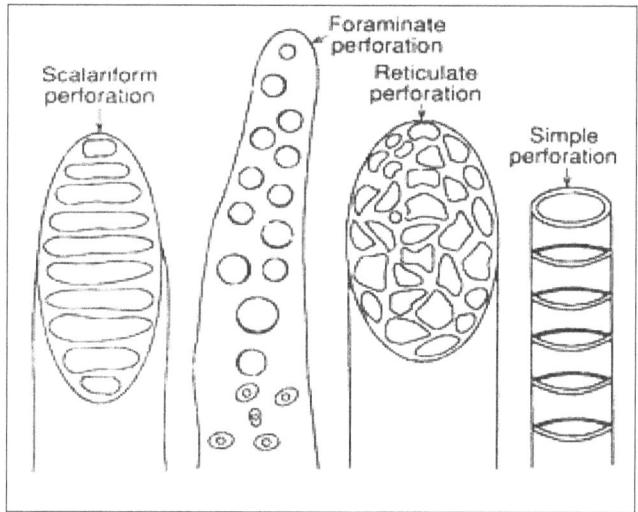

Different types of perforation plates found at the end walls of the vessles.

Distribution

Vessels predominate in the vascular tissues of angiosperms except some primitive groups like Ranales and Magnoliales as well as in Trochodendron and Tetracentron. They are absent in pteridophytes except in Selaginella, Equisetum, Pteridium etc. They are also absent in gymnosperms except Gnetales. They are present in the xylem of both primary and secondary bodies of angiosperms.

Origin and Phylogeny of Vessel

Vessels of primary xylem originate from procambium and that of secondary xylem develop from cambium. The vessels have evolved from usually long and narrow tracheids. So, the long vessels are primitive, while the short and wide vessels are advanced.

Vessels with long inclined ends are considered as primitive as it is present in tracheid-like vessels and those with transverse end walls are advanced. Scalariform perforation plate in a very inclined end wall of long vessel is considered as primitive.

Simple perforation with circular rim in an almost transverse end wall is the most advanced type. This type of vessel obviously facilitates the easy movement of water through it. The scalariform inclined perforation plates give resistance to water flow.

Scalariform pitting on the lateral wall of long vessel is most primitive whereas the alternate pitting on the lateral wall of short vessel is advanced.

Spiral thickening on the secondary wall, characteristic of the ring porous wood, is an advanced character.

Vessels' outlines in transverse sections have evolved in angular to circular direction. Angular outline is primitive than circular one.

It has been observed that the number of vessels present per square millimetre area of a transverse section of wood is a taxonomic character. High frequency of vessel is an advanced character.

Vessels may be present as single or in groups. These groupings and arrangements are taxonomic characters. The groups may be arranged in radial, oblique, or tangential lines. Solitary vessel represents the primitive condition.

Xylem Parenchyma

The parenchyma cells that occur as elements of the xylem tissue is termed as xylem parenchyma.

Distribution

Xylem parenchyma occurs in the primary and secondary xylem. In the latter they are present as axial and ray parenchyma. They are also known as wood parenchyma.

Structure and Arrangement

These cells may be oval, round, rectangular or square elongated and sometimes irregular in shape. The cell wall usually consists of thin primary wall. Sometimes the wall becomes thick due to lignin deposition over the primary cell wall. Pits when occur in-between two parenchyma cells are simple.

The pit pairs between the parenchyma and tracheary elements may be simple, half- bordered and bordered. Reserve foods in xylem parenchyma are starch and fat. Crystals and tannins are also found in these cells. The presence of chlorophyll is also reported in some herbs and deciduous trees.

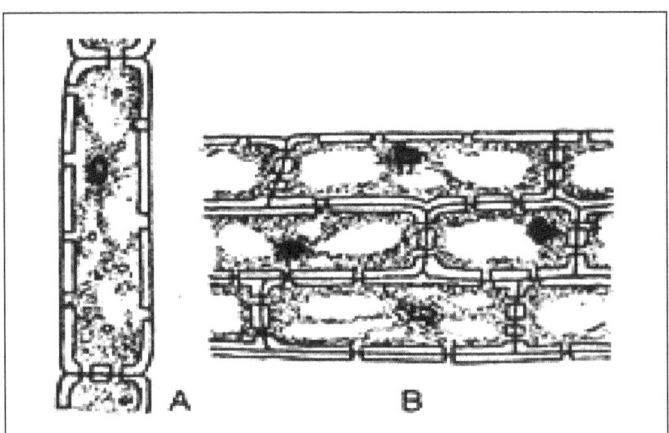

Xylem parenchyma: A. Parenchyma. B. Ray cells.

The xylem parenchyma cells are oriented vertically or horizontally. The vertical orientation forms parenchyma strands that are more common in secondary xylem. Sometimes parenchyma protrudes into vessels to form tylosis.

Function

Xylem parenchyma cell performs the following functions:

*   It helps in the transport of water and mineral salts.

*   It stores reserve food in the form of starch and fat and ergastic substances e.g., oils, gums, resin, tannins, silica bodies, crystals etc.

*   The thick walled lignified parenchyma gives mechanical support to some extent.

Ontogeny and Phylogeny

In primary xylem the xylem parenchyma originates from procambium. In secondary xylem the ray parenchyma cells originate from the ray initials of the cambium. The axial parenchyma along with tracheary elements and fibres originate from fusiform initials of cambium. The fusiform initials of cambium normally divide periclinally. In these cells anticlinal division also occur to keep pace with the growth of the stem in girth.

Phylogeny of Ray Parenchyma

Hetero- cellular multiseriate and uniseriate rays are the characteristics of the primitive wood. The multi- seriate rays have uniseriate wings. In advanced type of wood, however, either multiseriate or uniseriate ray is present (homocellular), that is one has been lost during the course of evolution.

The multiseriate rays are reduced in size and number. During evolution, the erect cells are lost and so there is a tendency for loss of heterogeneity. The advanced woods exhibit extremely short uniseriate wings on multiseriate rays.

In radial longitudinal section the ray parenchyma is observed to consist of square (isodiametric), erect (upright or vertical y elongate) and procumbent (radially elongate) cells. The square cells are morphologically equivalent to erect cells. The ray parenchyma may be homo- cellular and heterocellular.

In the former case rays are composed either of square cells or erect cells or procumbent cells, or erect and square cells' In the latter case, however, the rays consist of both square and procumbent cells or of erect and procumbent cells. The rays may be uniseriate or biseriate or multiseriate with uniseriate wings (tapered ends) as observed in tangential longitudinal section (TLS). The outlines of rays appear to be fusiform in TLS.

Phylogeny of Axial Parenchyma

The fusiform initial of cambium gives rise to axial (vertical) parenchyma. Axial parenchyma may occur independently or remain associated with vessels. So, depending on the presence of this parenchyma, the timbers may be apotracheal (vessels without axial parenchyma) and paratracheal (vessels with axial parenchyma).

The common apotracheal timbers may be:

- Diffuse (isolated strands of axial parenchyma),
- Diffuse-in-aggregates (axial parenchyma occurs as aggregates),
- Banded (axial parenchyma appears as narrow or wide bands), and
- Marginal (axial parenchyma occurs either at the beginning of growth ring or at the end).

The common paratracheal forms are:

- Scanty (discontinuous parenchymatous sheath surrounding a vessel),
- Vasicentric (continuous parenchymatous sheath surrounding the vessels),
- Abaxial (with more vasicentric parenchyma on abaxial side of vessel),
- Adaxial (with more parenchymatous width on adaxial side of vessel)
- Aliform (with winged vasicentric parenchyma), and
- Confluent (with continuous band of extended vasicentric parenchyma).

Phylogenetically, the primitive woods are usually devoid of parenchyma. If present, it is of diffuse type. The advanced wood exhibits a tendency towards grouping of axial parenchyma to form diffuse-in-aggregate type. The wide band of apotracheal parenchyma is more advanced than narrow band of parenchyma.

Xylem Fibre

The fibre components of xylem are known as xylary fibres or wood fibres. These fibres are of two types – libiriform fibre and fibre-tracheid. The libriform fibre is longer and thick- walled with simple pits.

Xylary fibres may be septate. In the tension wood the libiriform fibre and fibre tracheid are of gelatinous type. The xylary fibres are responsible for mechanical support. They may remain living for a longer period and store reserve food.

Phylogeny of Xylem Fibre

Fibres originate from procambium in primary xylem whereas that of secondary xylem develops from fusiform initial of cambium. Phylogenetically, the fibres have evolved from tracheids. During evolution the wall thickness has increased, the length has decreased and the bordered pits became reduced in size.

In dicotyledonous secondary xylem the evolution has occurred in the sequence of tracheid – fibre-tracheid – libiriform fibre. The tracheids have bordered pits; the pit border is diminished in fibre-tracheids and disappeared in libiriform fibres.

Phloem

Phloem is a complex vascular tissue through which photosynthates move from green tissues to the different parts of the plant Sometimes it adds mechanical strength to the plant body. It is one of the component tissues of the vascular bundle. It remains alternately arranged with the xylem (radial) in the roots. In the stem, however, the phloem is usually external to the xylem.

It may surround a central core of xylem (i.e., haplostele) or discrete strands of xylem remain surrounded by phloem (i.e., mixed protostele). Two cylinders of phloem may occur on both sides of xylem (i.e., amphiphloic siphonostele).

In some cases the central phloem strand may be encircled by xylem (i.e amphivasal, e.g., Dracaena). In most dicotyledonous stem phloem strand occurs external to xylem (e.g., collateral vascular bundle).

In some species of Cucurbitaceae, Asclepiadaceae, Apocynaceae, Solanaceae etc. there are both external and internal phloem tissues. They may also be termed as abaxial and adaxial phloem respectively. The internal phloem is also termed as intraxylary phloem. Phloem strands embedded in the secondary xylem is termed as included or interxylary phloem.

Phloem is mainly composed of:

- Parenchyma termed phloem parenchyma,

- Specialized parenchyma cells known as companion cells and albuminous cells,

- Phloem fibres,

- Sieve elements.

Sclereids, laticifers and resin ducts are also present in phloem tissues of some species. Phloem parenchyma, sieve tubes, companion cells and phloem fibres are the composition of phloem tissue in dicotyledonous plants. Phloem parenchyma is absent in monocots and a few members of Ranunculaceae. Sieve cells and albuminous cells are present in gymnosperm and vascular cryptogams.

Phloem Parenchyma

The parenchyma other than albuminous and companion cells in phloem is called phloem parenchyma.

Distribution

Phloem parenchyma is the component of both primary and secondary phloem. In secondary phloem they remain as axial phloem parenchyma and phloem rays.

Structure and Arrangement

Phloem parenchyma cells are rectangular or rounded in transverse section. It appears oblong with rounded or tapered ends in longitudinal section. The cell walls are thin and non-lignified with numerous pit fields through which plasmodesmatal connections are established with companion cells and sieve elements.

Sometimes sclerified and thick-walled parenchyma cells occur which are inactive. Phloem parenchyma cells with wall ingrowths are known as transfer cells. Phloem parenchyma may store starch, fats, resins, tannins etc.

In primary phloem the parenchyma cells remain parallel to the long axis of xylem, whereas, in secondary phloem, the axial parenchyma are parallel but the ray parenchyma are perpendicular to the long axis of the associated xylem.

Function

Phloem parenchyma cells perform the following functions:

- Phloem parenchyma cells perform the function of storage of fat, starch, resin tannins etc.

- Transfer cells are responsible for short distance transport of photosynthates.

- Phellogen derived from phloem parenchyma forms the periderm to protect the inner tissues.

- The inactive lignified cells add mechanical strength.

Phylogeny

Procambium gives rise to phloem parenchyma of primary phloem. In secondary phloem the axial phloem parenchyma and phloem rays are developed from fusiform initial and ray initial of cambium, respectively. Some parenchymal cells originate from a common mother cell of sieve elements. In many dicotyledonous stems the parenchyma of protophloem often differentiates into fibre in later stages of development, i.e., when the protophloem elements become functionless.

The phloem parenchyma and fibre of the secondary phloem bear no phylogenetic trend in phloem evolution. The distribution and morphology of them may be of comparative value.

Companion Cell

These cells with dense cytoplasm remain associated with the sieve tubes. Both of them originate from the same mother cell. Companion cells load photosynthates into the sieve tubes.

Structure and Arrangement

The cells are vertically elongated and somewhat angular in cross-section. They are usually shorter in length or may be as long as the associated sieve tubes. The uniformly thick cell wall possesses many sieve areas on the sieve tube side and primary pit fields on the opposite side. There are well- developed plasmodesmata in the sieve areas and primary pit fields. In some companion cells wall materials deposit on the inner side of the primary wall to form transfer cell.

The companion cells have prominent elongated or lobed nuclei. The cells abundantly contain Golgi apparatus, endoplasmic reticulum, mitochondria, ribosomes, plastids etc. In some companion cells P-proteins are found.

The sieve tube and companion cell remain strongly attached after their formation from the same mother cell to form a complex known as SE/CC complex. They cannot be separated by the usual maceration technique. Companion cells vary in number in relation to a single sieve tube.

Usually, the number is one or two and occasionally up to five or several. Companion cells are absent in primary phloem. They are also absent in some primitive woody dicotyledons, gymnosperms and pteridophytes. In gymnosperm the associated parenchyma with sieve cells are termed as albuminous cells.

Function

Following are the functions of the companion cell:

1. The companion cells are mainly related with the loading of the sieve tubes with sucrose. When the SE/CC complex is symplastically isolated the sucrose is absorbed actively by the companion cells from the apoplast of the minute veins.

From the companion cells the sugars then pass through the plasmodesmata into the sieve elements. But when there is protoplasmic continuity with the mesophyll cells through plasmodesmata sucrose is loaded through polymer trapping mechanism. The sieve tube becomes non-functional when the associated companion cell dies.

2. They are the active sites of protein synthesis.

Phylogeny

The companion cell and sieve tube have got the common origin. The mother cell naturally remains in the primary and secondary phloem. The parent cell itself originates either from procambium or from cambium respectively. The mother cell divides by unequal longitudinal division and the smaller cell differentiates into the companion cell.

Companion cells remain associated with sieve tubes.

Three Categories of Companion Cells are Observed

1. Companion cell may be much shorter than the accompanying sieve tube;
2. It may be almost as long as the sieve tube, and
3. It may be as long as the sieve tube, but septate.

According to many workers the first type is the primitive type. The remaining two types are advanced. Reduced number of companion cells in the phloem indicates the advancement.

Albuminous Cell

The parenchyma cells associated with the sieve cells are called albuminous cells. Each albuminous cell has a prominent nucleus and dense cytoplasm.

Structure and Arrangement

Albuminous cells are vertically oblong and may be of the same length as that of sieve cells or shorter.

There are symplastic connections between the two types of cells. The albuminous cells contain starch-free and protein-rich cytoplasm and stain deeply with cytoplasmic stains. These cells occur at the margins of rays.

Albuminous cells are present in both primary and secondary phloem. Therefore, their origin differs. In primary phloem they develop either from procambium derived phloem rays or from phloem parenchyma. In the secondary phloem of Ephedra, the albuminous cells originate from the fusiform initials of vascular cambium.

Function

There is a morphological as well as functional relationship between the albuminous cell and sieve cell even though they differ ontogenetically. When the sieve cell is non-functional the associated albuminous cell becomes dead. The possible function of the albuminous cell is to help in the conduction of proteins.

Phloem Fibre

The phloem fibres are the extraxylary fibres. They are also called bast fibres or bass fibre or basswood or bast wood fibres.

Structure and Arrangement

The fibres are elongated cells and may be very long with tapering ends interlocked with other fibres. Its thick cell wall is usually lignified. But in Linum phloem fibre wall is made of cellulose. Simple pits with linear or round apertures are found in the fibre wall. Slightly bordered pits may rarely occur. The phloem fibre of Vitis is septate.

The septate fibres store starch, oils, resins, calcium oxalate crystals etc. The fibres remain parallel to the long axis. In cross section they appear as isolated or scattered strands, as continuous or irregular bands, as clusters over the phloem strand and may form cylinders of tangential sheets encircling the inner tissues.

Function

The phloem fibres perform the following functions:

- The phloem fibres with their interlocked ends give mechanical strength to the plant.
- They protect the inner tissues.
- Septate fibres may store starch, oils, resins etc.

Phylogeny

The phloem fibres occur in both primary and secondary phloem and, therefore, their origins differ. The primary phloem fibre originates from procambium whereas the secondary phloem fibre originates from cambium. The fusiform initial of cambium gives rise to fibre.

Sieve Elements

The conducting elements of the phloem are referred to as sieve elements that are characterized by the presence of sieve areas and absence of nuclei from mature protoplasts. These include sieve cell and sieve tube. The sieve cells do not contain sieve plates. They are found in Pteridophytes and Gymnosperms.

Sieve Tube

It is the main solute conducting element in angiospermic phloem. It is living but enucleate and arranged longitudinally. Ribosomes and dictyosomes are also absent from mature protoplast. The sieve tubes have sieve plates and sieve areas on their transverse end walls. The sieve tube consists of longitudinally files of cells that are connected with each other through sieve areas on their transverse end walls.

The cell wall of the sieve tube may be thin or thick and is usually primary. The wall is composed mainly of cellulose and pectin. The sieve tubes of Smilax hispida, and Neptuma oleracea contain nucleus.

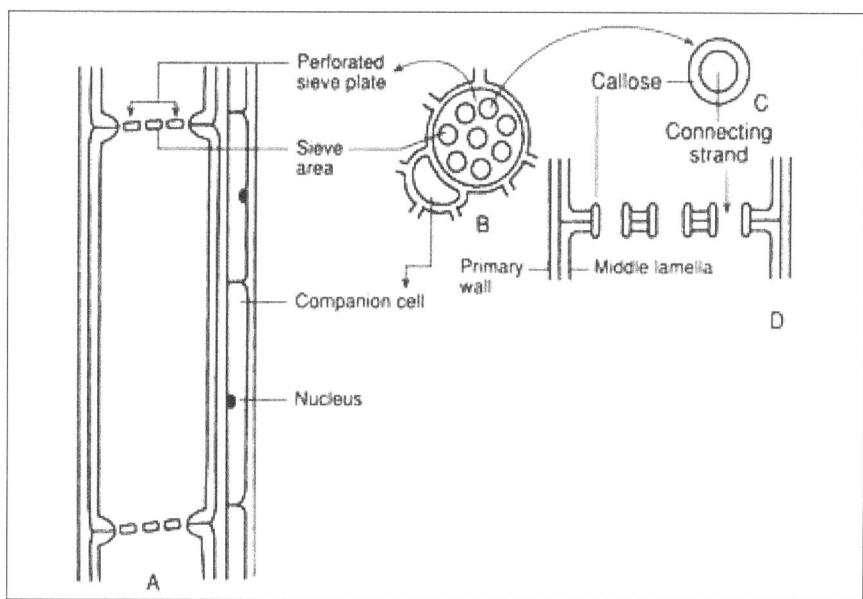

Diagrammatic representation of the sieve tube and companion cell in L.S.
(A) and T.S. (B). C. A single sieve area D. Vertical Section of sieve plate.

Table: Difference between albuminous cell and companion cell.

1.	It is present in the phloem of conifers and Ginkgo.	It is present in the phloem of angiosperm.
2.	Ontogenetically, it is not related to the sieve cell.	It is ontogenetically related to sieve tube.
3.	There is close morphological and physiological relationship with sieve cell.	Close morphological and physiological relationship exists with the sieve tubes.
4.	It is not so strongly attached with sieve cell.	It attaches strongly with the sieve tube and can-not separated after maceration.

The young sieve tube contains prominent nucleus, abundant dictyosomes, ribosomes, endoplasmic reticulum, plastids, mitochondria and other cell organelles. Sieve tubes actuate starch which stains brownish red with iodine instead of blue. Discrete proteinaceous slime bodies are found in the young sieve tubes in the form of filament, tubule, granule or crystal, called P-protein.

It occurs in all dicotyledonous species without any exception and is rare in monocotyledons. They are absent in gymnosperms excepting Ephedra and ptendophytes. The P-proteins are synthesised in the cytoplasm and occupies the peripheral position. In the stained preparation of sieve tubes P-proteins accumulate at the transverse end walls of the tubes and plug the sieve plate pores. This plug is termed slime plug.

Plastids occurring in the sieve tube protoplast may be either S-type or P-type depending on the nature of reserve food. Starch accumulating type is called S-type whereas protein accumulating type is called P-type plastid. S-type plastids are found in Bataceae, Polygonaceae Plumbaginaceae etc. The ultrastructure of the plastids of the sieve tubes is a taxonomic characteristic. These ultrastructural details of sieve element-plastids are, now-a-days, applied to characterise some higher taxa.

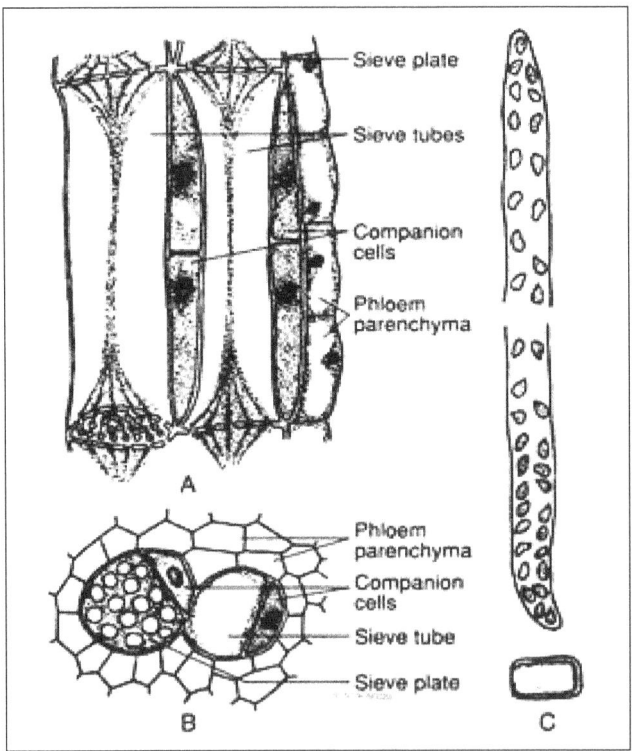

Sieve elements: A. Sieve tubes in L.S. B. Same in T.S.C. Sieve cell in L.S. and T.S.

The sieve plate is the region where the sieve areas occur. It is usually the horizontal or oblique end wall of the sieve tube. Sieve areas appear as depressed region on the wall where pores occur. Through the pores protoplasmic connections are established between the neighbouring members.

There may be one or several sieve areas in each sieve plate and accordingly they are termed as:

- Simple sieve plate with only one sieve area (e.g., Cucurbita), and

- Compound sieve plate with more than one sieve areas in the plate (e.g., Vitis, Pyrus etc.).

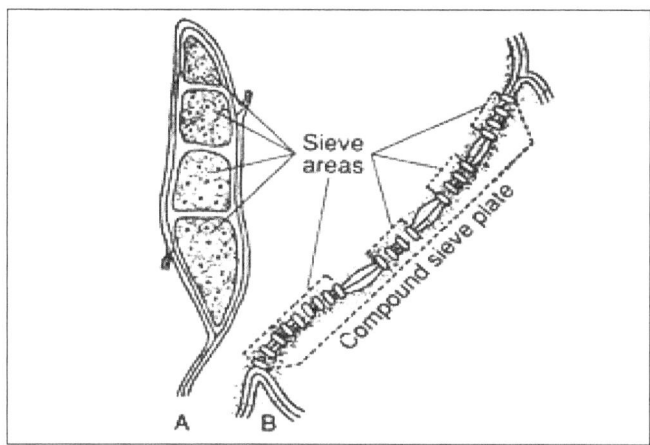

Sieve tube with compound sieve plate in surface view (A) and sectional view (B).

The sieve plate pores contain callose. It is a carbohydrate and is composed of β-1, 3-linked glucan. It sheaths the connecting strand in the pore. In mature sieve areas, callose deposition may be throughout the plate. With the increase in deposition the connecting strands in the pores gradually become thin and ultimately disappear.

The deposition of the neighbouring sieve areas may coalesce to a single mass and form the callose pad and, ultimately, the sieve tubes become non-functional. Callose deposition may be seasonal or permanent.

The former is usually referred to as dormancy callose (e.g. Vitis) and the latter as definitive callose. Callose is studied by staining it with aniline blue. When viewed with a microscope using ultraviolet light it fluoresces lemon-yellow colouration.

Sieve Cell

It is found in pteridophytes and gymnosperms. The sieve cells are arranged longitudinally, but, unlike sieve tubes, not one above the other in a series. The cells are oblong and tapered at their ends. They are often devoid of distinct end walls. When present, the end walls are tapered and oblique and may overlap.

The cell wall is usually thin and cellulosic but, in Pinus, the sieve cells are thick walled. Perforations on the walls form the sieve areas present on lateral walls and sometimes on the end walls. They are more numerous in the overlapping areas. Callose deposition in the perforations of sieve areas is also found. Sieve cells remain anucleate and living at maturity (exception: Pinus strobus and the family Taxaceae).

A large central vacuole is present pushing the protoplast towards the wall side forming the primordial utricle. Mitochondria, plastids and slime bodies are present. Starch grains are absent in sieve cells. In contrast to sieve tube, sieve cells are devoid of companion cell. They remain associated with albuminous cell.

Origin and Phylogeny of Sieve Tube

Sieve tubes in primary phloem originate from procambium and that of the secondary phloem originate from cambium. The mother cells divide longitudinally to form two daughter cells, one of which forms the companion cell and the other develops into sieve tube.

The differentiating sieve tube increases in length and, at maturity, it contains mitochondria, plastids, P-protein and endoplasmic reticulum. The nucleus disappears except in Taxus, Neptunia oleracea etc. Sieve areas originate at the sieve plate, the common transverse wall of the sieve tubes.

Phylogenetically, long sieve tubes with numerous sieve areas are considered as primitive whereas the shorter ones are advanced. The cambial initials, during the course of evolution tend to become shortened and this caused the formation of short sieve tubes.

Usually, the cambial initial undergoes transverse septation before the differentiation of sieve tube leading to the formation of short sieve tubes. Small pores in the sieve plate are regarded as primitive. The simple horizontal sieve plate with single sieve area is an advanced feature.

Origin and Phylogeny of Sieve Cell

The mother cells of sieve cells are slender, short cylindrical to oblong with tapering ends and numerous primary pit fields on their lateral walls. During differentiation, the mother cells elongate and the cell wall becomes thick. Sieve areas develop in the position of pit-fields. Plasmodesmata appear in the sieve areas and callose develops surrounding them.

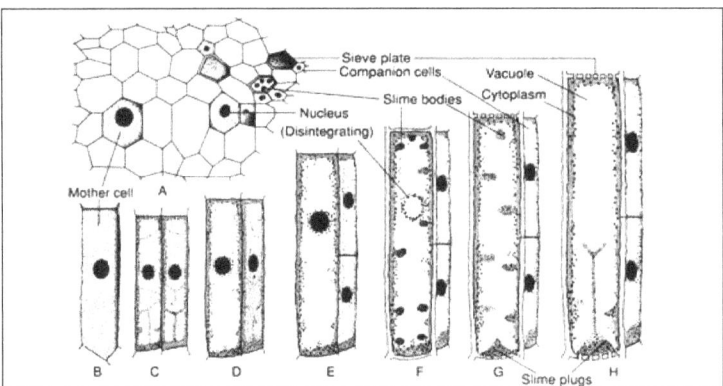

Differential of sieve tube: A. T.S through phloem showing stages. B-H, L.S through phloem showing different stages.

Sieve cell predominates in pteridophytes and gymnosperms. It is reported among angiosperms in Austrobaileya scandens and Sorbus aucuparla. Sieve cells are considered as primitive. They are not arranged in axial files. Moreover, the structure of end walls is similar to the lateral walls.

Special Tissues

The group of cells usually of different origin concerned in secretion of oils, nectar, resins, gums, mucilage, latex etc. is generally termed as secretory tissues. This categorisation is based on physiological functions. The secretory cells may be any part of the plant body as a solitary cell, or group of cells, or a whole tissue like pith, cortex, xylem phloem etc.

Secretion means the process of release of by-products of metabolism from a cell. The secretory substances may also be stored in insoluble forms within the cell. These substances have either a special physiological function or may be secreted or stored as wastes.

The removal of waste products of metabolism is defined as excretion. Strictly, the secretory products take part in the metabolism like hormones and enzymes. At the same time the process of cell wall deposition, cutinisation, wax deposition, suberisation etc. are also the examples of secretion. On the other hand, examples of excretory products are terpenes, saponins, rubber, tannins, crystals etc. Of course, there is no sharp demarcation between the two terms secretion and excretion.

The secretory bodies vary greatly in structure and position. They are mainly of two types:

- The secretory cells in which the secretory substances are formed and exuded cell outside e.g., simple glandular trichromes, nectaries, multicellular glands with vascular tissues etc., and

- The structures in which the secretory substances are stored and released after the breakdown of the cells. The secretory structures may be external and internal.

Merismatic Tissue

A meristem is the tissue in most plants containing undifferentiated cells (meristematic cells), found in zones of the plant where growth can take place. Meristematic cells give rise to various organs of a plant and are responsible for growth.

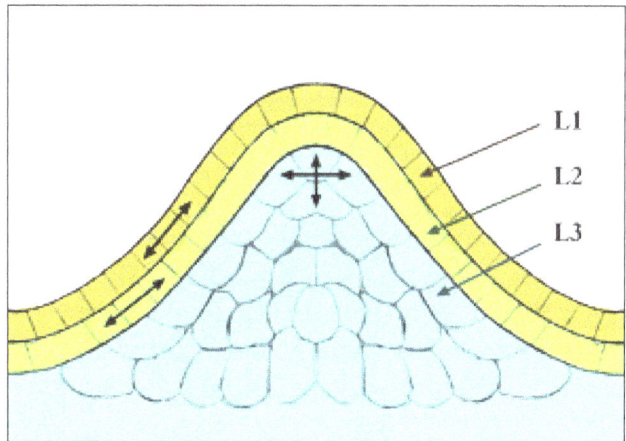

Tunica-Corpus model of the apical meristem (growing tip). The epidermal (L1) and subepidermal (L2) layers form the outer layers called the tunica. The inner L3 layer is called the corpus. Cells in the L1 and L2 layers divide in a sideways fashion, which keeps these layers distinct, whereas the L3 layer divides in a more random fashion.

Differentiated plant cells generally cannot divide or produce cells of a different type. Meristematic cells are incompletely or not at all differentiated, and are capable of continued cellular division. Therefore, cell division in the meristem is required to provide new cells for expansion and differentiation of tissues and initiation of new organs, providing the basic structure of the plant body. Furthermore, the cells are small and protoplasm fills the cell completely. The vacuoles are extremely small. The cytoplasm does not contain differentiated plastids (chloroplasts or chromoplasts), although they are present in rudimentary form (proplastids). Meristematic cells are packed closely together without intercellular cavities. The cell wall is a very thin primary cell wall as well as some are thick in some plants. Maintenance of the cells requires a balance between two antagonistic processes: organ initiation and stem cell population renewal.

There are three types of meristematic tissues: apical (at the tips), intercalary (in the middle) and lateral (at the sides). At the meristem summit, there is a small group of slowly dividing cells, which is commonly called the central zone. Cells of this zone have a stem cell function and are essential for meristem maintenance. The proliferation and growth rates at the meristem summit usually differ considerably from those at the periphery.

Apical Meristems

Apical meristems are the completely undifferentiated (indeterminate) meristems in a plant. These differentiate into three kinds of primary meristems. The primary meristems in turn produce the two secondary meristem types. These secondary meristems are also known as lateral meristems because they are involved in lateral growth.

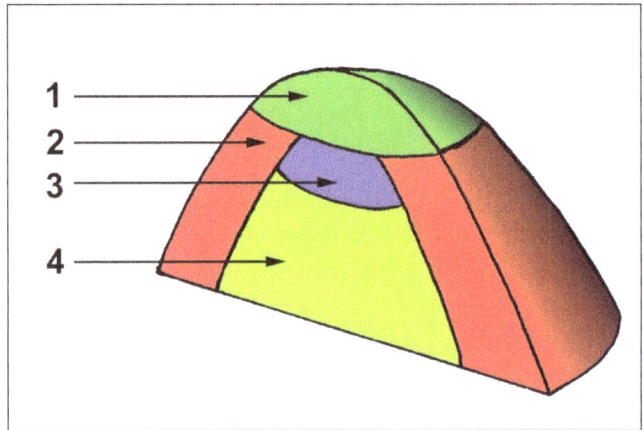

Organisation of an apical meristem (growing tip) 1 - Central zone 2 - Peripheral zone
3 - Medullary (i.e. central) meristem 4 - Medullary tissue.

There are two types of apical meristem tissue: shoot apical meristem (SAM), which gives rise to organs like the leaves and flowers, and root apical meristem (RAM), which provides the meristematic cells for future root growth. SAM and RAM cells divide rapidly and are considered indeterminate, in that they do not possess any defined end status. In that sense, the meristematic cells are frequently compared to the stem cells in animals, which have an analogous behavior and function.

The number of layers varies according to plant type. In general the outermost layer is called the tunica while the innermost layers are the corpus. In monocots, the tunica determine the physical characteristics of the leaf edge and margin. In dicots, layer two of the corpus determine the characteristics of the edge of the leaf. The corpus and tunica play a critical part of the plant physical appearance as all plant cells are formed from the meristems. Apical meristems are found in two locations: the root and the stem. Some Arctic plants have an apical meristem in the lower/middle parts of the plant. It is thought that this kind of meristem evolved because it is advantageous in Arctic conditions.

Shoot Apical Meristems

Shoot apical meristems are the source of all above-ground organs, such as leaves and flowers. Cells at the shoot apical meristem summit serve as stem cells to the surrounding peripheral region, where they proliferate rapidly and are incorporated into differentiating leaf or flower primordia.

Shoot apical meristems of Crassula ovata (left). Fourteen days later, leaves have developed (right).

The shoot apical meristem is the site of most of the embryogenesis in flowering plants. Primordia of leaves, sepals, petals, stamens, and ovaries are initiated here at the rate of one every time interval, called a plastochron. It is where the first indications that flower development has been evoked are manifested. One of these indications might be the loss of apical dominance and the release of otherwise dormant cells to develop as auxiliary shoot meristems, in some species in axils of primordia as close as two or three away from the apical dome. The shoot apical meristem consists of 4 distinct cell groups:

- Stem cells,

- The immediate daughter cells of the stem cells,

- A subjacent organizing center,

- Founder cells for organ initiation in surrounding regions.

The four distinct zones mentioned above are maintained by a complex signalling pathway. In Arabidopsis thaliana, 3 interacting CLAVATA genes are required to regulate the size of the stem cell reservoir in the shoot apical meristem by controlling the rate of cell division. CLV1 and CLV2 are predicted to form a receptor complex (of the LRR receptor-like kinase family) to which CLV3 is a ligand. CLV3 shares some homology with the ESR proteins of maize, with a short 14 amino acid region being conserved between the proteins. Proteins that contain these conserved regions have been grouped into the CLE family of proteins.

CLV1 has been shown to interact with several cytoplasmic proteins that are most likely involved in downstream signalling. For example, the CLV complex has been found to be associated with Rho/Rac small GTPase-related proteins. These proteins may act as an intermediate between the CLV complex and a mitogen-activated protein kinase (MAPK), which is often involved in signalling cascades. KAPP is a kinase-associated protein phosphatase that has been shown to interact with CLV1. KAPP is thought to act as a negative regulator of CLV1 by dephosphorylating it.

Another important gene in plant meristem maintenance is WUSCHEL (shortened to WUS), which is a target of CLV signaling in addition to positively regulating CLV, thus forming a feedback loop. WUS is expressed in the cells below the stem cells of the meristem and its presence prevents the differentiation of the stem cells. CLV1 acts to promote cellular differentiation by repressing WUS activity outside of the central zone containing the stem cells.

The function of WUS in the shoot apical meristem is linked to the phytohormone cytokinin. Cytokinin activates histidine kinases which then phosphorylate histidine phosphotransfer proteins. Subsequently, the phosphate groups are transferred onto two types of Arabidopsis response regulators (ARRs): Type-B ARRS and Type-A ARRs. Type-B ARRs work as transcription factors to activate genes downstream of cytokinin, including A-ARRs. A-ARRs are similar to B-ARRs in structure; however, A-ARRs do not contain the DNA binding domains that B-ARRs have, and which are required to function as transcription factors. Therefore, A-ARRs do not contribute to the activation of transcription, and by competing for phosphates from phosphotransfer proteins, inhibit B-ARRs function. In the SAM, B-ARRs induce the expression of WUS which induces stem cell identity. WUS then suppresses A-ARRs. As a result, B-ARRs are no longer inhibited, causing sustained cytokinin signaling in the center of the shoot apical meristem. Altogether with CLAVATA signaling, this system works as a negative feedback loop. Cytokinin signaling is positively reinforced by WUS to prevent the inhibition of cytokinin signaling, while WUS promotes its own inhibitor in the form of CLV3, which ultimately keeps WUS and cytokinin signaling in check.

Root Apical Meristem

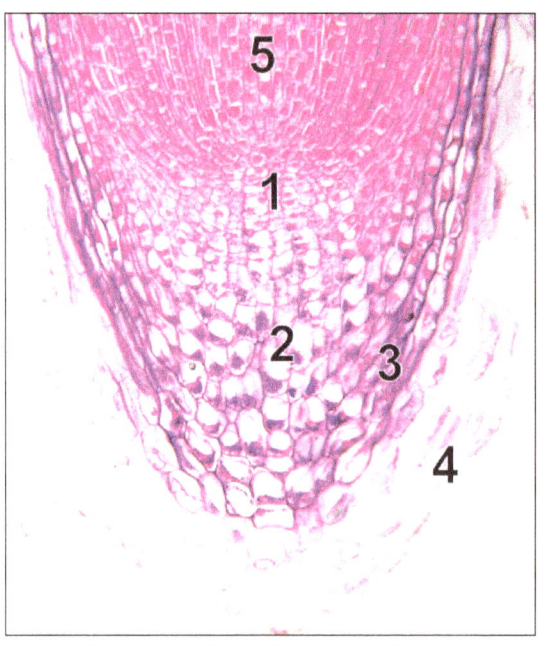

10× microscope image of root tip with meristem. 1 - quiescent center 2 - calyptrogen (live rootcap cells) 3 - rootcap 4 - sloughed off dead rootcap cells 5 – procambium.

Unlike the shoot apical meristem, the root apical meristem produces cells in two dimensions. It harbors two pools of stem cells around an organizing center called the quiescent center (QC) cells and together produces most of the cells in an adult root. At its apex, the root meristem is covered by the root cap, which protects and guides its growth trajectory. Cells are continuously sloughed off the outer surface of the root cap. The QC cells are characterized by their low mitotic activity. Evidence suggests that the QC maintains the surrounding stem cells by preventing their differentiation, via signal(s) that are yet to be discovered. This allows a constant supply of new cells in the meristem required for continuous root growth. Recent findings indicate that QC can also act as a reservoir of stem cells to replenish whatever is lost or damaged. Root apical meristem and tissue patterns become established in the embryo in the case of the primary root, and in the new lateral root primordium in the case of secondary roots.

Intercalary Meristem

In angiosperms, intercalary meristems occur only in monocot (in particular, grass) stems at the base of nodes and leaf blades. Horsetails also exhibit intercalary growth. Intercalary meristems are capable of cell division, and they allow for rapid growth and regrowth of many monocots. Intercalary meristems at the nodes of bamboo allow for rapid stem elongation, while those at the base of most grass leaf blades allow damaged leaves to rapidly regrow. This leaf regrowth in grasses evolved in response to damage by grazing herbivores.

Floral Meristem

When plants begin flowering, the shoot apical meristem is transformed into an inflorescence meristem, which goes on to produce the floral meristem, which produces the sepals, petals, stamens, and carpels of the flower.

In contrast to vegetative apical meristems and some efflorescence meristems, floral meristems cannot continue to grow indefinitely. Their future growth is limited to the flower with a particular size and form. The transition from shoot meristem to floral meristem requires floral meristem identity genes, that both specify the floral organs and cause the termination of the production of stem cells. AGAMOUS (AG) is a floral homeotic gene required for floral meristem termination and necessary for proper development of the stamens and carpels. AG is necessary to prevent the conversion of floral meristems to inflorescence shoot meristems, but is identity gene LEAFY (LFY) and WUS and is restricted to the centre of the floral meristem or the inner two whorls. This way floral identity and region specificity is achieved. WUS activates AG by binding to a consensus sequence in the AG's second intron and LFY binds to adjacent recognition sites. Once AG is activated it represses expression of WUS leading to the termination of the meristem.

Through the years, scientists have manipulated floral meristems for economic reasons. An example is the mutant tobacco plant "Maryland Mammoth." In 1936, the department of agriculture of Switzerland performed several scientific tests with this plant. "Maryland Mammoth" is peculiar in that it grows much faster than other tobacco plants.

Apical Dominance

Apical dominance is the phenomenon where one meristem prevents or inhibits the growth of other meristems. As a result, the plant will have one clearly defined main trunk. For example, in trees, the tip of the main trunk bears the dominant shoot meristem. Therefore, the tip of the trunk grows rapidly and is not shadowed by branches. If the dominant meristem is cut off, one or more branch tips will assume dominance. The branch will start growing faster and the new growth will be vertical. Over the years, the branch may begin to look more and more like an extension of the main trunk. Often several branches will exhibit this behavior after the removal of apical meristem, leading to a bushy growth.

The mechanism of apical dominance is based on auxins, types of plant growth regulators. These are produced in the apical meristem and transported towards the roots in the cambium. If apical dominance is complete, they prevent any branches from forming as long as the apical meristem is active. If the dominance is incomplete, side branches will develop.

Recent investigations into apical dominance and the control of branching have revealed a new plant hormone family termed strigolactones. These compounds were previously known to be involved in seed germination and communication with mycorrhizal fungi and are now shown to be involved in inhibition of branching.

Diversity in Meristem Architectures

The SAM contains a population of stem cells that also produce the lateral meristems while the stem elongates. It turns out that the mechanism of regulation of the stem cell number might be evolutionarily conserved. The CLAVATA gene CLV2 responsible for maintaining the stem cell population in Arabidopsis thaliana is very closely related to the Maize gene FASCIATED EAR 2(FEA2) also involved in the same function. Similarly, in Rice, the FON1-FON2 system seems to bear a close relationship with the CLV signaling system in Arabidopsis thaliana. These studies suggest that the regulation of stem cell number, identity and differentiation might be an evolutionarily conserved mechanism in monocots, if not in angiosperms. Rice also contains another genetic system distinct from FON1-FON2, that is involved in regulating stem cell number. This example underlines the innovation that goes about in the living world all the time.

Role of the KNOX-Family Genes

Complex leaves of Cardamine hirsuta result from KNOX gene expression.

Genetic screens have identified genes belonging to the KNOX family in this function. These genes essentially maintain the stem cells in an undifferentiated state. The KNOX family has undergone quite a bit of evolutionary diversification while keeping the overall mechanism more or less similar. Members of the KNOX family have been found in plants as diverse as Arabidopsis thaliana, rice, barley and tomato. KNOX-like genes are also present in some algae, mosses, ferns and gymnosperms. Misexpression of these genes leads to the formation of interesting morphological features. For example, among members of Antirrhineae, only the species of the genus Antirrhinum lack a structure called spur in the floral region. A spur is considered an evolutionary innovation because it defines pollinator specificity and attraction. Researchers carried out transposon mutagenesis in Antirrhinum majus, and saw that some insertions led to formation of spurs that were very similar to the other members of Antirrhineae, indicating that the loss of spur in wild Antirrhinum majus populations could probably be an evolutionary innovation.

Note the long spur of the above flower. Spurs attract pollinators and confer pollinator specificity.

The KNOX family has also been implicated in leaf shape evolution. One study looked at the pattern of KNOX gene expression in A. thaliana, that has simple leaves and Cardamine hirsuta, a plant having complex leaves. In A. thaliana, the KNOX genes are completely turned off in leaves, but in C.hirsuta, the expression continued, generating complex leaves. Also, it has been proposed that the mechanism of KNOX gene action is conserved across all vascular plants, because there is a tight correlation between KNOX expression and a complex leaf morphology.

Primary Meristems

Apical meristems may differentiate into three kinds of primary meristem:

- Protoderm: lies around the outside of the stem and develops into the epidermis.

- Procambium: lies just inside of the protoderm and develops into primary xylem and primary phloem. It also produces the vascular cambium, and cork cambium, secondary meristems. The cork cambium further differentiates into the phelloderm (to the inside) and the phellem, or cork (to the outside). All three of these layers (cork cambium, phellem, and phelloderm) constitute the periderm. In roots, the procambium can also give rise to the pericycle, which produces lateral roots in eudicots.

- Ground meristem: develops into the cortex and the pith. Composed of parenchyma, collenchyma and sclerenchyma cells.

These meristems are responsible for primary growth, or an increase in length or height, which were discovered by scientist Joseph D. Carr of North Carolina in 1943.

Secondary Meristems

There are two types of secondary meristems, these are also called the lateral meristems because they surround the established stem of a plant and cause it to grow laterally (i.e., larger in diameter).

- Vascular cambium, which produces secondary xylem and secondary phloem. This is a process that may continue throughout the life of the plant. This is what gives rise to wood in plants. Such plants are called arborescent. This does not occur in plants that do not go through secondary growth (known as herbaceous plants).

- Cork cambium, which gives rise to the periderm, which replaces the epidermis.

Indeterminate Growth of Meristems

Though each plant grows according to a certain set of rules, each new root and shoot meristem can go on growing for as long as it is alive. In many plants, meristematic growth is potentially indeterminate, making the overall shape of the plant not determinate in advance. This is the primary growth. Primary growth leads to lengthening of the plant body and organ formation. All plant organs arise ultimately from cell divisions in the apical meristems, followed by cell expansion and differentiation. Primary growth gives rise to the apical part of many plants.

The growth of nitrogen-fixing root nodules on legume plants such as soybean and pea is either determinate or indeterminate. Thus, soybean (or bean and Lotus japonicus) produce determinate nodules (spherical), with a branched vascular system surrounding the central infected zone. Often, Rhizobium infected cells have only small vacuoles. In contrast, nodules on pea, clovers, and Medicago truncatula are indeterminate, to maintain (at least for some time) an active meristem that yields new cells for Rhizobium infection. Thus zones of maturity exist in the nodule. Infected cells usually possess a large vacuole. The plant vascular system is branched and peripheral.

Cloning

Under appropriate conditions, each shoot meristem can develop into a complete, new plant or clone. Such new plants can be grown from shoot cuttings that contain an apical meristem. Root apical meristems are not readily cloned, however. This cloning is called asexual reproduction or vegetative reproduction and is widely practiced in horticulture to mass-produce plants of a desirable genotype. This process is also known as mericloning.

Propagating through cuttings is another form of vegetative propagation that initiates root or shoot production from secondary meristematic cambial cells. This explains why basal 'wounding' of shoot-borne cuttings often aids root formation.

Induced Meristems

Meristems may also be induced in the roots of legumes such as soybean, Lotus japonicus, pea, and Medicago truncatula after infection with soil bacteria commonly called Rhizobia. Cells of the inner or outer cortex in the so-called "window of nodulation" just behind the developing root tip are induced to divide. The critical signal substance is the lipo-oligosaccharide Nod factor, decorated with side groups to allow specificity of interaction. The Nod factor receptor proteins NFR1 and NFR5 were cloned from several legumes including Lotus japonicus, Medicago truncatula and soybean (Glycine max). Regulation of nodule meristems utilizes long-distance regulation known as the autoregulation of nodulation (AON). This process involves a leaf-vascular tissue located LRR receptor kinases (LjHAR1, GmNARK and MtSUNN), CLE peptide signalling, and KAPP interaction, similar to that seen in the CLV1,2,3 system. LjKLAVIER also exhibits a nodule regulation phenotype though it is not yet known how this relates to the other AON receptor kinases.

Vascular Tissue

Vascular tissue is a complex conducting tissue, formed of more than one cell type, found in vascular plants. The primary components of vascular tissue are the xylem and phloem. These two tissues transport fluid and nutrients internally. There are also two meristems associated with vascular tissue: the vascular cambium and the cork cambium. All the vascular tissues within a particular plant together constitute the vascular tissue system of that plant.

The cells in vascular tissue are typically long and slender. Since the xylem and phloem function in the conduction of water, minerals, and nutrients throughout the plant, it is not surprising that their form should be similar to pipes. The individual cells of phloem are connected end-to-end, just as the sections of a pipe might be. As the plant grows, new vascular tissue differentiates in the growing tips of the plant. The new tissue is aligned with existing vascular tissue, maintaining its connection throughout the plant. The vascular tissue in plants is arranged in long, discrete strands called vascular bundles. These bundles include both xylem and phloem, as well as supporting and protective cells. In stems and roots, the xylem typically lies closer to the interior of the stem with phloem towards the exterior of the stem. In the stems of some Asterales dicots, there may be phloem located inwardly from the xylem as well.

Cross section of celery stalk, showing vascular bundles, which include both phloem and xylem.

Between the xylem and phloem is a meristem called the vascular cambium. This tissue divides off cells that will become additional xylem and phloem. This growth increases the girth of the plant, rather than its length. As long as the vascular cambium continues to produce new cells, the plant will continue to grow more stout. In trees and other plants that develop wood, the vascular cambium allows the expansion of vascular tissue that produces woody growth. Because this growth ruptures the epidermis of the stem, woody plants also have a cork cambium that develops among the phloem. The cork cambium gives rise to thickened cork cells to protect the surface of the plant and reduce water loss. Both the production of wood and the production of cork are forms of secondary growth.

In leaves, the vascular bundles are located among the spongy mesophyll. The xylem is oriented toward the adaxial surface of the leaf (usually the upper side), and phloem is oriented toward the abaxial surface of the leaf. This is why aphids are typically found on the undersides of the leaves rather than on the top, since the phloem transports sugars manufactured by the plant and they are closer to the lower surface.

Detail of the vasculature of a bramble leaf.

Vascular Bundle

A vascular bundle is a part of the transport system in vascular plants. The transport itself happens in vascular tissue, which exists in two forms: xylem and phloem. Both these tissues are present in a vascular bundle, which in addition will include supporting and protective tissues.

The xylem typically lies adaxial with phloem positioned abaxial. In a stem or root this means that the xylem is closer to the centre of the stem or root while the phloem is closer to the exterior. In a leaf, the adaxial surface of the leaf will usually be the upper side, with the abaxial surface the lower side. This is why aphids are typically found on the underside of a leaf rather than on the top, since the sugars manufactured by the plant are transported by the phloem, which is closer to the lower surface.

The position of vascular bundles relative to each other may vary considerably:

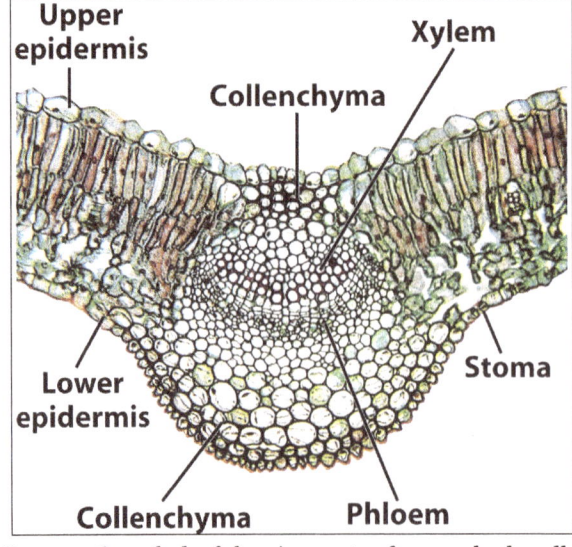

Cross section of a leaf showing parts of a vascular bundle.

Bundle-sheath Cells

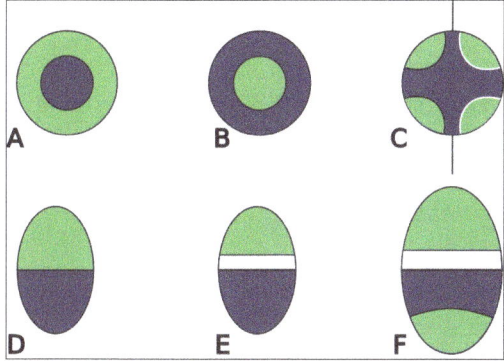

Types of Vascular bundles (blue: Xylem, green: Phloem, white: Cambium) A concentric, periphloematic B concentric, perixylematic C radial with inner xylem, here with four xylem-poles, left closed, right open D collateral closed E collateral open F bicollateral open.

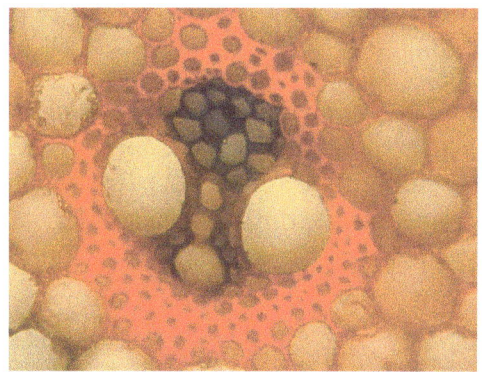

Detail of vascular bundle: closed, collateral vascular bundles of the stem axis of Zea mays.

Bundle-sheath cells are photosynthetic cells arranged into tightly packed sheaths around the veins of a leaf. They form a protective covering on leaf veins, and consist of one or more cell layers, usually parenchyma. Loosely arranged mesophyll cells lie between the bundle sheath and the leaf surface. The Calvin cycle is confined to the chloroplasts of these bundle sheath cells in C_4 plants.

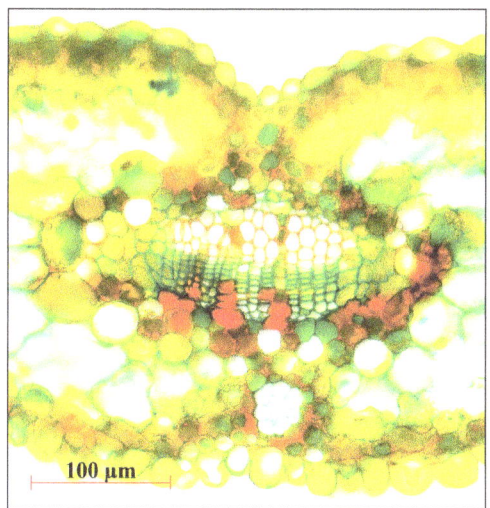

100 μm

Vascular bundle in the leaf of Metasequoia glyptostroboides.

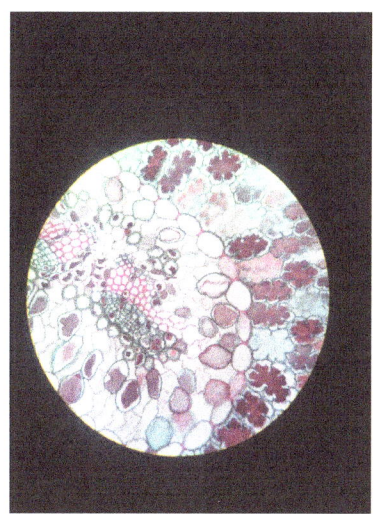

The vascular bundle of pine leaf showing xylem and phloem.

Vascular Cambium

The vascular cambium is the main growth tissue in the stems and roots of many plants, specifically in dicots such as buttercups and oak trees, gymnosperms such as pine trees, as well as in certain vascular plants. It produces secondary xylem inwards, towards the pith, and secondary phloem outwards, towards the bark. In herbaceous plants, it occurs in the vascular bundles which are often arranged like beads on a necklace forming an interrupted ring inside the stem. In woody plants, it forms a cylinder of unspecialized meristem cells, as a continuous ring from which the new tissues are grown. Unlike the xylem and phloem, it does not transport water, minerals or food through the plant. Other names for the vascular cambium are the main cambium, wood cambium, or bifacial cambium.

Vascular cambia are found in dicots and gymnosperms but not monocots, which usually lack secondary growth. A few leaf types also have a vascular cambium. In dicot and gymnosperm trees, the vascular cambium is the obvious line separating the bark and wood; they also have a cork cambium. For successful grafting, the vascular cambia of the rootstock and scion must be aligned so they can grow together.

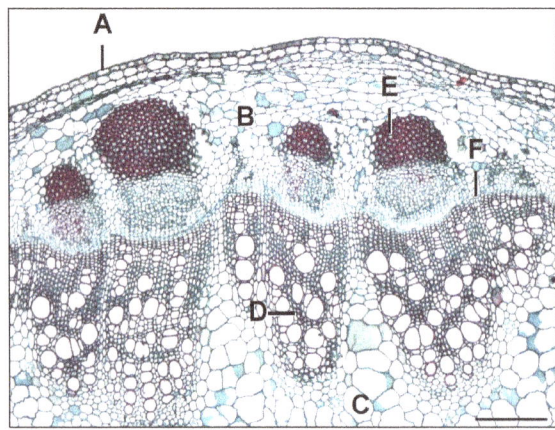

Helianthus stem in section. The cells of the vascular cambium (F) divide to form phloem on the outside, seen located beneath the bundle cap (E), and xylem (D) on the inside. Most of the vascular cambium is here in vascular bundles (ovals of phloem and xylem together) but it is starting to join these up as at point F between the bundles.

Structure and Function

The cambium present between primary xylem and primary phloem is called the intrafascicular cambium (within vascular bundles). During secondary growth, cells of medullary rays, in a line (as seen in section; in three dimensions, it is a sheet) between neighbouring vascular bundles, become meristematic and form new interfascicular cambium (between vascular bundles). The intrafascicular and interfascicular cambia thus join up to form a ring (in three dimensions, a tube) which separates the primary xylem and primary phloem, the cambium ring. The vascular cambium produces secondary xylem on the inside of the ring, and secondary phloem on the outside, pushing the primary xylem and phloem apart.

The vascular cambium usually consists of two types of cells:

- Fusiform initials (tall, axially oriented).

- Ray initials (smaller and round to angular in shape).

Maintenance of Cambial Meristem

The vascular cambium is maintained by a network of interacting signal feedback loops. Currently, both hormones and short peptides have been identified as information carriers in these systems. While similar regulation occurs in other plant meristems, the cambial meristem receives signals from both the xylem and phloem sides for the meristem. Signals received from outside the meristem act to down regulate internal factors, which promotes cell proliferation and differentiation.

Hormonal Regulation

The phytohormones that are involved in the vascular cambial activity are auxins, ethylene, gibberellins, cytokinins, abscisic acid and probably more to be discovered. Each one of these plant hormones are vital for regulation of cambial activity. Combination of different concentrations of these hormones is very important in plant metabolism.

Auxin hormones are proven to stimulate mitosis, cell production and regulate interfascicular and fascicular cambium. Applying auxin to the surface of a tree stump allowed decapitated shoots to continue secondary growth. The absence of auxin hormones will have a detrimental effect on a plant. It has been shown that mutants without auxin will exhibit increased spacing between the interfascicular cambiums and reduced growth of the vascular bundles. The mutant plant will therefore experience a decreased in water, nutrients, and photosynthates being transported throughout the plant, eventually leading to death. Auxin also regulates the two types of cell in the vascular cambium, ray and fusiform initials. Regulation of these initials ensures the connection and communication between xylem and phloem is maintained for the translocation of nourishment and sugars are safely being stored as an energy resource. Ethylene levels are high in plants with an active cambial zone and are still currently being studied. Gibberellin stimulates the cambial cell division and also regulates differentiation of the xylem tissues, with no effect on the rate of phloem differentiation. Differentiation is an essential process that changes these tissues into a more specialized type, leading to an important role in maintaining the life form of a plant. In poplar trees, high concentrations of gibberellin is positively correlated to an increase of cambial cell division and an increase of auxin in the cambial stem cells. Gibberellin is also responsible for the expansion of xylem through a signal traveling from the shoot to the root. Cytokinin hormone is known to regulate the rate of the cell division instead of the direction of cell differentiation. A study demonstrated that the mutants are found to have a reduction in stem and root growth but the secondary vascular pattern of the vascular bundles were not affected with a treatment of cytokinin.

Cork Cambium

Cork cambium (pl. cambia or cambiums) is a tissue found in many vascular plants as a part of the epidermis. It is one of the many layers of bark, between the cork and primary phloem. The cork cambium is a lateral meristem and is responsible for secondary growth that replaces the epidermis in roots and stems. It is found in woody and many herbaceous dicots, gymnosperms and some monocots (monocots usually lack secondary growth). It is one of the plant's meristems – the series of tissues consisting of embryonic disk (incompletely differentiated) cells from which the plant grows. The function of cork cambium is to produce the cork, a tough protective material.

Synonyms for cork cambium are bark cambium, pericambium and phellogen. Phellogen is defined as the meristematic cell layer responsible for the development of the periderm. Cells that grow inwards from there are termed phelloderm, and cells that develop outwards are termed phellem or cork. The periderm thus consists of three different layers:

- Phelloderm – inside of cork cambium; composed of living parenchyma cells,

- Phellogen (cork cambium) – meristem that gives rise to periderm,

- Phellem (cork) – dead at maturity; air-filled protective tissue on the outside.

Growth and development of cork cambium is very variable between different species, and is also highly dependent on age and growth conditions, as can be observed from the different surfaces of bark, which may be smooth, fissured, tesselated, scaly, or flaking off.

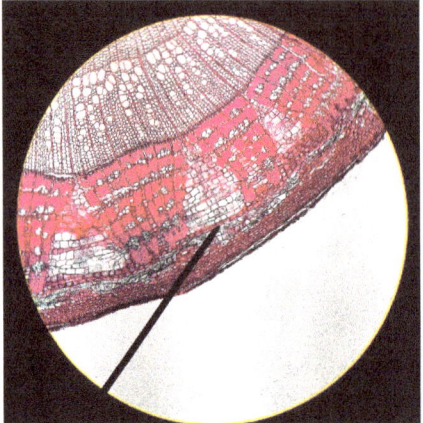

Cork cambium of woody stem (Tilia). It is different from the main vascular cambium, which is the ring between the wood (xylem) on the inside (top) and the red bast (phloem) outside it.

Quercus suber (cork oak) bark, Portugal.

Pith

Pith, or medulla, is a tissue in the stems of vascular plants. Pith is composed of soft, spongy parenchyma cells, which store and transport nutrients throughout the plant. In eudicotyledons, pith is located in the center of the stem. In monocotyledons, it extends also into flowering stems and roots. The pith is encircled by a ring of xylem; the xylem, in turn, is encircled by a ring of phloem.

While new pith growth is usually white or pale in colour, as the tissue ages it commonly darkens to a deeper brown color. In trees pith is generally present in young growth, but in the trunk and older branches the pith often gets replaced - in great part - by xylem. In some plants, the pith in the middle of the stem may dry out and disintegrate, resulting in a hollow stem. A few plants, such as walnuts, have distinctive chambered pith with numerous short cavities. The cells in the peripheral parts of the pith may, in some plants, develop to be different from cells in the rest of the pith. This layer of cells is then called the perimedullary region of the pithamus. An example of this can be observed in Hedera helix, a species of ivy.

Elder shoot cut longitudinally to show the broad, solid pith (rough-textured, white) inside the wood (smooth, yellow-tinged). Scale in mm.

The term pith is also used to refer to the pale, spongy inner layer of the rind, more properly called mesocarp or albedo, of citrus fruits (such as oranges) and other hesperidia. The pith of the sola or other similar plants is used to make the pith helmet.

Walnut shoot cut longitudinally to show the chambered pith found in this genus. Scale in mm.

The tiny centre dark spot (about 1 mm diameter) in this yew wood is the pith.

The pith of the sago palm, although highly toxic to animals in its raw form, is an important human food source in Melanesia and Micronesia by virtue of its starch content and its availability. There is a simple process of starch extraction from sago pith that leaches away a sufficient amount of the toxins and thus only the starch component is consumed. Current processes for starch extraction are generally only about 50% efficient, however, with the other half remaining in residual pith waste. The form of the starch after processing is similar to tapioca.

Ground Tissue

The ground tissue of plants includes all tissues that are neither dermal nor vascular. It can be divided into three types based on the nature of the cell walls.

- Parenchyma cells have thin primary walls and usually remain alive after they become mature. Parenchyma forms the "filler" tissue in the soft parts of plants, and is usually present in cortex, pericycle, pith, and medullary rays in primary stem and root.

- Collenchyma cells have thin primary walls with some areas of secondary thickening. Collenchyma provides extra mechanical and structural support, particularly in regions of new growth.

- Sclerenchyma cells have thick lignified secondary walls and often die when mature. Sclerenchyma provides the main structural support to a plant.

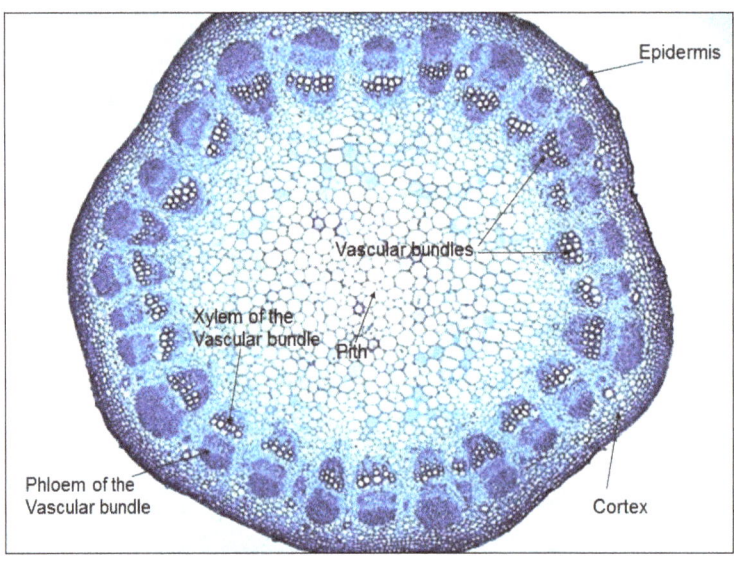

Parenchyma

Parenchyma is a versatile ground tissue that generally constitutes the "filler" tissue in soft parts of plants. It forms, among other things, the cortex (outer region) and pith (central region) of stems, the cortex of roots, the mesophyll of leaves, the pulp of fruits, and the endosperm of seeds. Parenchyma cells are living cells and may remain meristematic at maturity—meaning that they are capable of cell division if stimulated. They have thin and flexible cellulose cell walls, and are generally polyhedral when close-packed, but can be roughly spherical when isolated from their neighbours. Parenchyma cells are generally large. They have large central vacuoles, which allow the cells to store and regulate ions, waste products, and water. Tissue specialised for food storage is commonly formed of parenchyma cells.

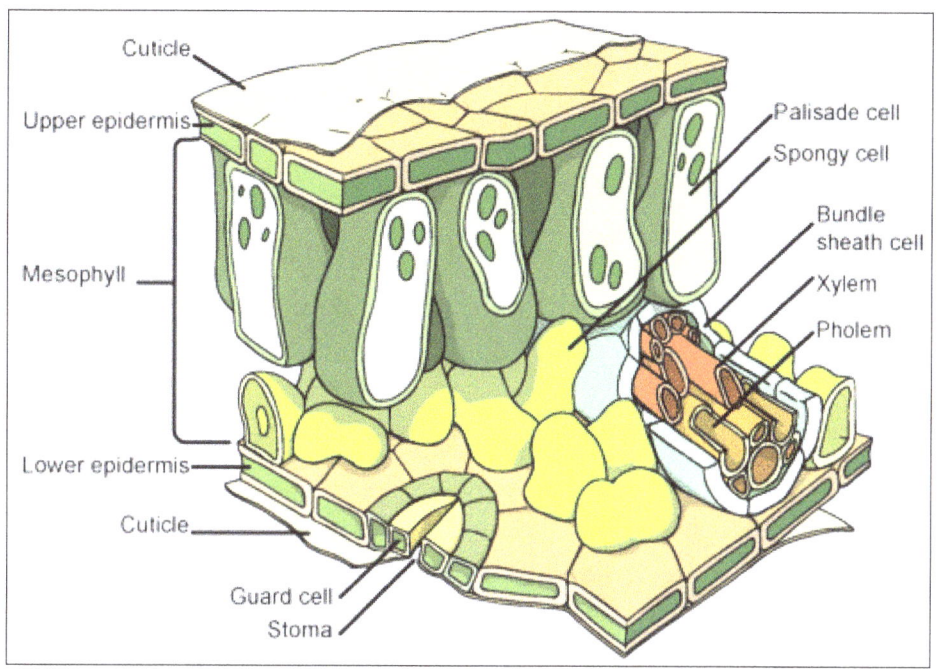

Cross section of a leaf showing various ground tissue types.

Parenchyma cells have a variety of functions:

- Their main function is to repair.

- In leaves, they form two layers of mesophyll cells immediately beneath the epidermis of the leaf, that are responsible for photosynthesis and the exchange of gases. These layers are called the palisade parenchyma and spongy mesophyll. Palisade parenchyma cells can be either cuboidal or elongated. Parenchyma cells in the mesophyll of leaves are specialised parenchyma cells called chlorenchyma cells (parenchyma cells with chloroplasts). Chlorenchyma cells are also found in other parts of the plant.

- Storage of starch, protein, fats, oils and water in roots, tubers (e.g. potatoes), seed endosperm (e.g. cereals) and cotyledons (e.g. pulses and peanuts).

- Secretion (e.g. the parenchyma cells lining the inside of resin ducts).

- Wound repair and the potential for renewed meristematic activity.

- Other specialised functions such as aeration (aerenchyma) provides buoyancy and helps aquatic plants float.

- Chlorenchyma cells carry out photosynthesis and manufacture food.

The shape of parenchyma cells varies with their function. In the spongy mesophyll of a leaf, parenchyma cells range from near-spherical and loosely arranged with large intercellular spaces, to branched or stellate, mutually interconnected with their neighbours at the ends of their arms to form a three-dimensional network, like in the red kidney bean Phaseolus vulgaris and other mesophytes. These cells, along with the epidermal guard cells of the stoma, form a system of air spaces and chambers that regulate the exchange of gases. In some works, the cells of the leaf epidermis are regarded as specialised parenchymal cells, but the modern preference has long been to classify the epidermis as plant dermal tissue, and parenchyma as ground tissue.

Shapes of parenchyma:

- Polyhedral (found in pallisade tissue of the leaf).

- Circular.

- Stellate (found in stem of plants and have well developed air spaces between them).

- Elongated (also found in pallisade tissue of leaf).

- Lobed (found in spongy and pallisade mesophyyll tissue of some plants).

Collenchyma

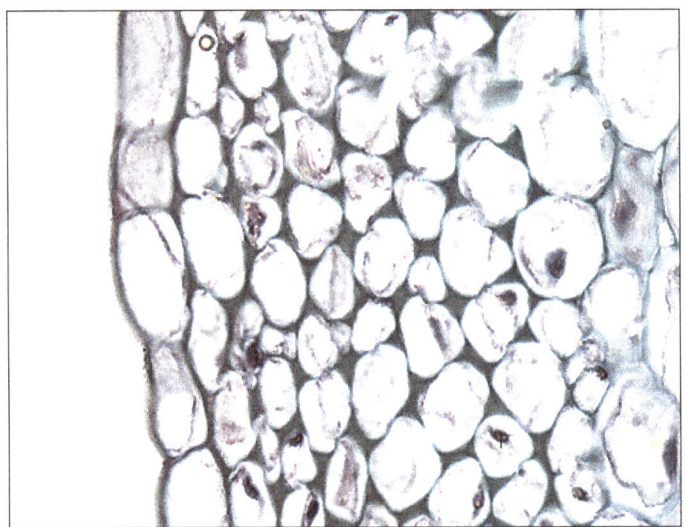

Cross section of collenchyma cells.

Collenchyma tissue is composed of elongated cells with irregularly thickened walls. They provide structural support, particularly in growing shoots and leaves. Collenchyma tissue makes up things such as the resilient strands in stalks of celery. Collenchyma cells are usually living, and have only a thick primary cell wall made up of cellulose and pectin. Cell wall thickness is strongly affected by mechanical stress upon the plant. The walls of collenchyma in shaken plants (to mimic the effects of wind etc.), may be 40–100% thicker than those not shaken.

There are four main types of collenchyma:

- Angular collenchyma (thickened at intercellular contact points).

- Tangential collenchyma (cells arranged into ordered rows and thickened at the tangential face of the cell wall).

- Annular collenchyma (uniformly thickened cell walls).

- Lacunar collenchyma (collenchyma with intercellular spaces).

Collenchyma cells are most often found adjacent to outer growing tissues such as the vascular cambium and are known for increasing structural support and integrity.

The first use of "collenchyma" was by Link who used it to describe the sticky substance on Bletia (Orchidaceae) pollen. Complaining about Link's excessive nomenclature, Schleiden stated mockingly that the term "collenchyma" could have more easily been used to describe elongated sub-epidermal cells with unevenly thickened cell walls.

Sclerenchyma

Sclerenchyma is the tissue which makes the plant hard and stiff. Sclerenchyma is the supporting tissue in plants. Two types of sclerenchyma cells exist: fibers and sclereids. Their cell walls consist of cellulose, hemicellulose, and lignin. Sclerenchyma cells are the principal supporting cells in plant tissues that have ceased elongation. Sclerenchyma fibers are of great economic importance, since they constitute the source material for many fabrics (e.g. [flax] hemp, jute, and ramie).

Unlike the collenchyma, mature sclerenchyma is composed of dead cells with extremely thick cell walls (secondary walls) that make up to 90% of the whole cell volume. It is the hard, thick walls that make sclerenchyma cells important strengthening and supporting elements in plant parts that have ceased elongation. The difference between fibers and sclereids is not always clear: transitions do exist, sometimes even within the same plant.

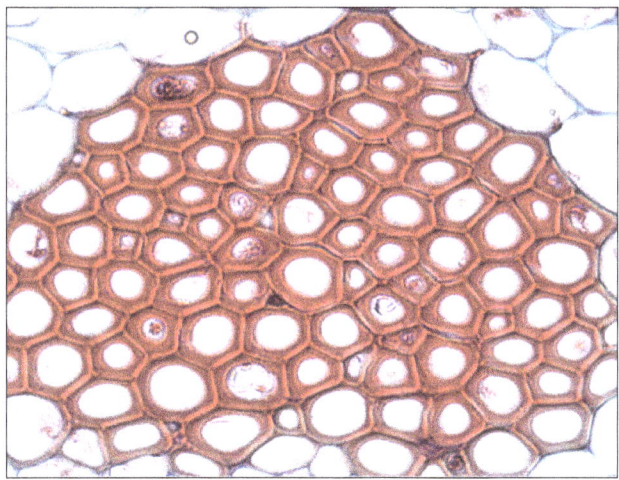

Cross section of sclerenchyma fibers.

Fibers or bast are generally long, slender, so-called prosenchymatous cells, usually occurring in strands or bundles. Such bundles or the totality of a stem's bundles are colloquially called fibers. Their high load-bearing capacity and the ease with which they can be processed has since antiquity made them the source material for a number of things, like ropes, fabrics and mattresses. The fibers of flax (Linum usitatissimum) have been known in Europe and Egypt for more than 3,000 years, those of hemp (Cannabis sativa) in China for just as long. These fibers, and those of jute (Corchorus capsularis) and ramie (Boehmeria nivea, a nettle), are extremely soft and elastic and are especially well suited for the processing to textiles. Their principal cell wall material is cellulose.

Contrasting are hard fibers that are mostly found in monocots. Typical examples are the fiber of many grasses, agaves (sisal: Agave sisalana), lilies (Yucca or Phormium tenax), Musa textilis and others. Their cell walls contain, besides cellulose, a high proportion of lignin. The load-bearing capacity of Phormium tenax is as high as 20–25 kg/mm², the same as that of good steel wire (25 kg/ mm²), but the fibre tears as soon as too great a strain is placed upon it, while the wire distorts and does not tear before a strain of 80 kg/mm². The thickening of a cell wall has been studied in Linum. Starting at the centre of the fiber, the thickening layers of the secondary wall are deposited one after the other. Growth at both tips of the cell leads to simultaneous elongation. During development the layers of secondary material seem like tubes, of which the outer one is always longer and older than the next. After completion of growth, the missing parts are supplemented, so that the wall is evenly thickened up to the tips of the fibers.

Fibers usually originate from meristematic tissues. Cambium and procambium are their main centers of production. They are usually associated with the xylem and phloem of the vascular bundles. The fibers of the xylem are always lignified, while those of the phloem are cellulosic. Reliable evidence for the fibre cells' evolutionary origin from tracheids exists. During evolution the strength of the tracheid cell walls was enhanced, the ability to conduct water was lost and the size of the pits was reduced. Fibers that do not belong to the xylem are bast (outside the ring of cambium) and such fibers that are arranged in characteristic patterns at different sites of the shoot. The term "sclerenchyma" (originally Sclerenchyma) was introduced by Mettenius in 1865.

Sclereids

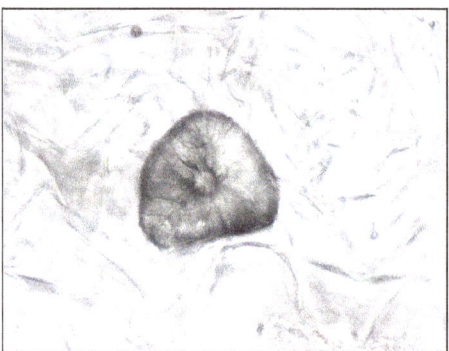

Fresh mount of a sclereid.

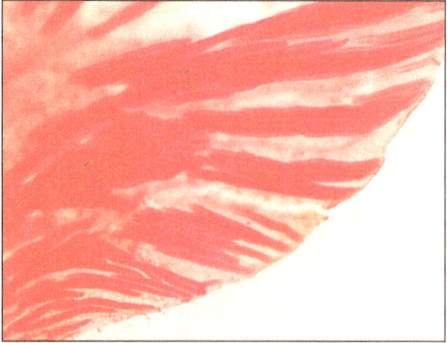

Long, tapered sclereids supporting a leaf edge in Dionysia kossinskyi.

Sclereids are the reduced form of sclerenchyma cells with highly thickened, lignified walls.

They are small bundles of sclerenchyma tissue in plants that form durable layers, such as the cores of apples and the gritty texture of pears (Pyrus communis). Sclereids are variable in shape. The cells can be isodiametric, prosenchymatic, forked or elaborately branched. They can be grouped into bundles, can form complete tubes located at the periphery or can occur as single cells or small groups of cells within parenchyma tissues. But compared with most fibres, sclereids are relatively short. Characteristic examples are brachysclereids or the stone cells (called stone cells because of their hardness) of pears and quinces (Cydonia oblonga) and those of the shoot of the wax plant (Hoya carnosa). The cell walls fill nearly all the cell's volume. A layering of the walls and the existence of branched pits is clearly visible. Branched pits such as these are called ramiform pits. The shell of many seeds like those of nuts as well as the stones of drupes like cherries and plums are made up from sclereids.

Dermal Tissue

The dermal tissue is composed of the epidermis, the outermost layer that covers plant body as a protective envelope. It is the skin that covers leaves, flowers, roots, fruits and seeds.

In typical cases, consists of a layer of cells which are joining without spaces, forming a continuous film, while allowing the exchange of substances with the environment. These cells have a cell wall cellulose not very thickened.

The cytoplasm is sparse and is retracted toward the cell wall by large vacuoles, which may have pigments or lack thereof. They have a core of small size compared with the volume of the cell. The plastids are represented by small leucoplasts or missing entirely.

Some plants have chloroplasts in the epidermis like ferns and aquatic plants of shady habitat.

The epidermis can have stomatal, an openings through which plant exchanges gas with the atmosphere. These openings are surrounded by specialized cells called occlusive which change size and shape, modifying the diameter of the stomatal opening and thereby regulate the gas exchange.

The epidermis is coated with a film of wax called the cuticle, is waterproof, and its function is to reduce evaporative water loss through the plant surface. If it undergoes secondary growth rather than epidermis will have periderm, the tissue composed of cells almost completely waterproof (especially corky tissue or cork) who die when mature.

Epidermis

The epidermis is a single layer of cells that covers the leaves, flowers, roots and stems of plants. It forms a boundary between the plant and the external environment. The epidermis serves several functions: it protects against water loss, regulates gas exchange, secretes metabolic compounds, and (especially in roots) absorbs water and mineral nutrients. The epidermis of most leaves shows dorsoventral anatomy: the upper (adaxial) and lower (abaxial) surfaces have somewhat different construction and may serve different functions. Woody stems and some other stem structures such as potato tubers produce a secondary covering called the periderm that replaces the epidermis as the protective covering.

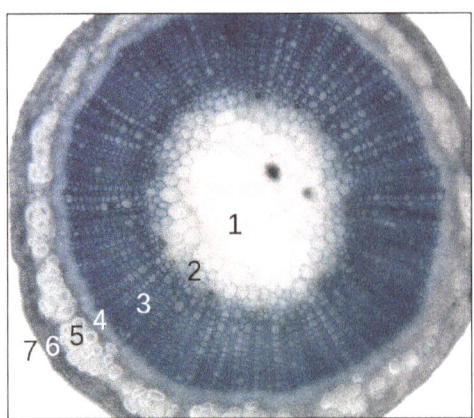

Cross-section of a flax plant stem: 1. pith 2. protoxylem 3. xylem 4. phloem
5. sclerenchyma (bast fibre) 6. cortex 7. epidermis.

The epidermis is the outermost cell layer of the primary plant body. In some older works the cells of the leaf epidermis have been regarded as specialised parenchyma cells, but the established modern preference has long been to classify the epidermis as dermal tissue, whereas parenchyma is classified as ground tissue. The epidermis is the main component of the dermal tissue system of leaves, and also stems, roots, flowers, fruits, and seeds; it is usually transparent (epidermal cells have fewer chloroplasts or lack them completely, except for the guard cells).

The cells of the epidermis are structurally and functionally variable. Most plants have an epidermis that is a single cell layer thick. Some plants like Ficus elastica and Peperomia, which have periclinal cellular division within the protoderm of the leaves, have an epidermis with multiple cell layers. Epidermal cells are tightly linked to each other and provide mechanical strength and protection to the plant. The walls of the epidermal cells of the above ground parts of plants contain cutin, and are covered with a cuticle. The cuticle reduces water loss to the atmosphere, it is sometimes covered with wax in smooth sheets, granules, plates, tubes or filaments. The wax layers give some plants a whitish or bluish surface color. Surface wax acts as a moisture barrier and protects the plant from intense sunlight and wind. The underside of many leaves have a thinner cuticle than the top side, and leaves of plants from dry climates often have thickened cuticles to conserve water by reducing transpiration.

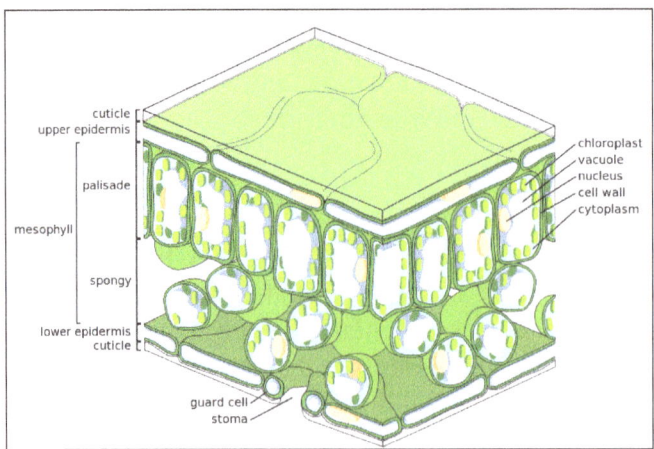

The epidermal tissue includes several differentiated cell types: epidermal cells, guard cells, subsidiary cells, and epidermal hairs (trichomes). The epidermal cells are the most numerous, largest, and least specialized. These are typically more elongated in the leaves of monocots than in those of dicots.

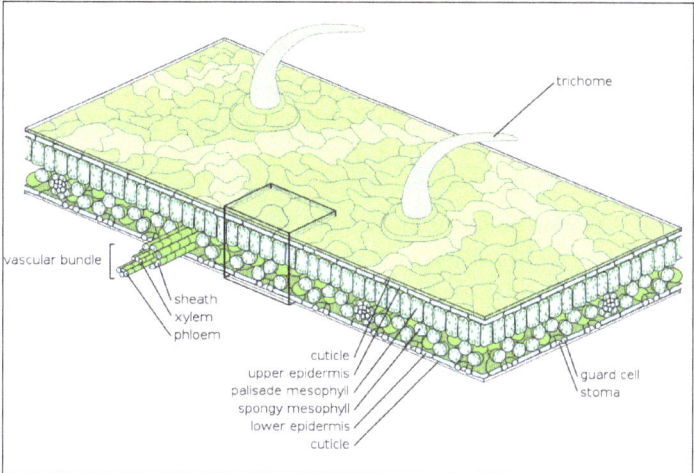

Trichomes or hairs grow out from the epidermis in many species. In root epidermis, epidermal hairs, termed root hairs are common and are specialized for absorption of water and mineral nutrients.

In plants with secondary growth, the epidermis of roots and stems is usually replaced by a periderm through the action of a cork cambium.

Guard Cells

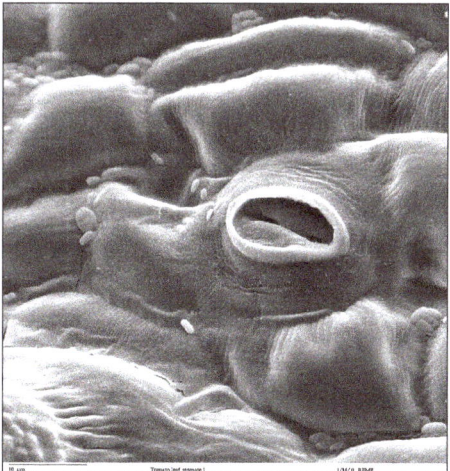

Stoma in a tomato leaf.

The leaf and stem epidermis is covered with pores called stomata (sing., stoma), part of a stoma complex consisting of a pore surrounded on each side by chloroplast-containing guard cells, and two to four subsidiary cells that lack chloroplasts. The stomata complex regulates the exchange of gases and water vapor between the outside air and the interior of the leaf. Typically, the stomata are more numerous over the abaxial (lower) epidermis of the leaf than the (adaxial) upper epidermis. An exception is floating leaves where most or all stomata are on the upper surface. Vertical leaves, such as those of many grasses, often have roughly equal numbers of stomata on both surfaces. The stoma is bounded by two guard cells. The guard cells differ from the epidermal cells in the following aspects:

- The guard cells are bean-shaped in surface view, while the epidermal cells are irregular in shape.

- The guard cells contain chloroplasts, so they can manufacture food by photosynthesis (The epidermal cells do not contain chloroplasts).

- Guard cells are the only epidermal cells that can make sugar. According to one theory, in sunlight the concentration of potassium ions (K+) increases in the guard cells. This, together with the sugars formed, lowers the water potential in the guard cells. As a result, water from other cells enter the guard cells by osmosis so they swell and become turgid. Because the guard cells have a thicker cellulose wall on one side of the cell, i.e. the side around the stomatal pore, the swollen guard cells become curved and pull the stomata open.

At night, the sugar is used up and water leaves the guard cells, so they become flaccid and the stomatal pore closes. In this way, they reduce the amount of water vapour escaping from the leaf.

Cell Differentiation in the Epidermis

The plant epidermis consists of three main cell types: pavement cells, guard cells and their subsidiary cells that surround the stomata and trichomes, otherwise known as leaf hairs. The epidermis of petals also form a variation of trichomes called conical cells.

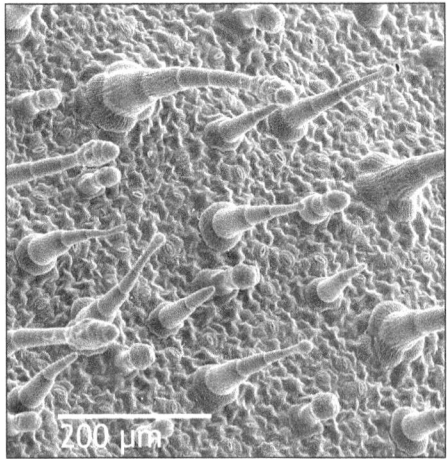

Scanning electron microscope image of Nicotiana alata leaf's epidermis, showing trichomes
(hair-like appendages) and stomata (eye-shaped slits, visible at full resolution).

Trichomes develop at a distinct phase during leaf development, under the control of two major trichome specification genes: TTG and GL1. The process may be controlled by the plant hormones gibberellins, and even if not completely controlled, gibberellins certainly have an effect on the development of the leaf hairs. GL1 causes endoreplication, the replication of DNA without subsequent cell division as well as cell expansion. GL1 turns on the expression of a second gene for trichome formation, GL2, which controls the final stages of trichome formation causing the cellular outgrowth.

Arabidopsis thaliana uses the products of inhibitory genes to control the patterning of trichomes, such as TTG and TRY. The products of these genes will diffuse into the lateral cells, preventing them from forming trichomes and in the case of TRY promoting the formation of pavement cells.

Expression of the gene MIXTA, or its analogue in other species, later in the process of cellular differentiation will cause the formation of conical cells over trichomes. MIXTA is a transcription factor.

Stomatal patterning is a much more controlled process, as the stoma effect the plants water retention and respiration capabilities. As a consequence of these important functions, differentiation of cells to form stomata is also subject to environmental conditions to a much greater degree than other epidermal cell types.

Stomata are pores in the plant epidermis that are surrounded by two guard cells, which control the opening and closing of the aperture. These guard cells are in turn surrounded by subsidiary cells which provide a supporting role for the guard cells.

Stomata begin as stomatal meristemoids. The process varies between dicots and monocots. Spacing is thought to be essentially random in dicots though mutants do show it is under some form of genetic control, but it is more controlled in monocots, where stomata arise from specific asymmetric divisions of protoderm cells. The smaller of the two cells produced becomes the guard mother cells. Adjacent epidermal cells will also divide asymmetrically to form the subsidiary cells.

Because stomata play such an important role in the plants survival, collecting information on their differentiation is difficult by the traditional means of genetic manipulation, as stomatal mutants tend to be unable to survive. Thus the control of the process is not well understood. Some genes have been identified. TMM is thought to control the timing of stomatal initiation specification and FLP is thought to be involved in preventing further division of the guard cells once they are formed.

Environmental conditions affect the development of stomata, in particular their density on the leaf surface. It is thought that plant hormones, such as ethylene and cytokines, control the stomatal developmental response to the environmental conditions. Accumulation of these hormones appears to cause increased stomatal density such as when the plants are kept in closed environments.

Secretory Plant Tissue

The tissues that are concerned with the secretion of gums, resins, volatile oils, nectar latex, and other substances in plants are called secretory tissues. These tissues are classified as either laticiferous tissues or glandular tissues.

Cells or organizations of cells which produce a variety of secretions. The secreted substance may remain deposited within the secretory cell itself or may be excreted, that is, released from the cell. Substances may be excreted to the surface of the plant or into intercellular cavities or canals. Some of the many substances contained in the secretions are not further utilized by the plant (resins, rubber, tannins, and various crystals), while others take part in the functions of the plant (enzymes and hormones). Secretory structures range from single cells scattered among other kinds of cells to complex structures involving many cells; the latter are often called glands.

Epidermal hairs of many plants are secretory or glandular. Such hairs commonly have a head composed of one or more secretory cells borne on a stalk. The hair of a stinging needle is bulbous below and extends into a long, fine process above. If one touches the hair, its tip breaks off, the sharp edge penetrates the skin, and the poisonous secretion is released. Glands secreting a sugary liquid—the nectar—in flowers pollinated by insects are called nectaries. Nectaries may occur on the floral stalk or on any floral organ: sepal, petal, stamen, or ovary.

The hydathode structures discharge water—a phenomenon called guttation through openings in margins or tips of leaves. The water flows through the xylem to its endings in the leaf and then through the intercellular spaces of the hydathode tissue toward the openings in the epidermis. Strictly speaking, such hydathodes are not glands because they are passive with regard to the flow of water.

Some carnivorous plants have glands that produce secretions capable of digesting insects and small animals. These glands occur on leaf parts modified as insect-trapping structures. In the sundews (Drosera) the traps bear stalked glands, called tentacles. When an insect lights on the leaf, the tentacles bend down and cover the victim with a mucilaginous secretion, the enzymes of which digest the insect.

Resin ducts are canals lined with secretory cells that release resins into the canal. Resin ducts are common in gymnosperms and occur in various tissues of roots, stems, leaves, and reproductive structures.

Gum ducts are similar to resin ducts and may contain resins, oils, and gums. Usually, the term gum duct is used with reference to the dicotyledons, although gum ducts also may occur in the gymnosperms. Oil ducts are intercellular canals whose secretory cells produce oils or similar substances. Such ducts may be seen, for example, in various parts of the plant of the carrot family (Umbelliferae). Laticifers are cells or systems of cells containing latex, a milky or clear, colored or colorless liquid. Latex occurs under pressure and exudes from the plant when the latter is cut.

Laticiferous Tissues

These consist of thick walled, greatly elongated an d much branched ducts containing a milky or yellowish colored juice known as latex. They contain numerous nuclei which lie embedded in the thin lining layer of protoplasm. They irregularly distributed in the mass of parenchymatous cells. Laticiferous ducts, in which latex are found are again two types:

- Latex cell or non-articulate latex ducts.

- Latex vessels or articulate latex.

Latex Cells

Also called as "non-articulate latex ducts", these ducts are independent units which extend as branched structures for long distances in the plant body. They originates as minute structures, elongate quickly and by repeated branching ramify in all directions but do not fuse together. Thus a network is not formed as in latex vessels.

Latex Vessel

Also called "articulate latex ducts", these ducts or vessels are the result of anastamosis of many cells. They grow more or less as parallel ducts which by means of branching and frequent anastamoses form a complex network. Latex vessels are commonly found in many angiosperm families Papaveraceae, Compositae, Euphorbiaceae, Moraceae, etc.

Function

The function of laticiferous ducts is not clearly understood. They may also act as food storage organs or as reservoir of waste products, or as translocatory tissues.

Glandular Tissues

This tissue consists of special structures; the glands. These glands contain some secretory or excretory products. A gland may consist of isolated cells or small group cells with or without a central cavity. They are of various kinds and may be internal or external.

Internal glands are:

- Oil-gland secreting essential oils, as in the fruits and leaves of orange, lemon.
- Mucilage secreting glands, as in the betel leaf.
- Glands secreting gum, resin, tannin, etc.
- Digestive glands secreting enzymes or digestive agents.
- Special water secreting glands at the tip of veins.

External glands are commonly short hairs tipped by glands. They are:

- water-secreting hairs or glands.
- Glandular hairs secreting gum like substances as in tobacco, plumbago, etc.
- Glandular hairs secreting irritating, poisonous substances, as in nettles.
- Honey glands, as in carnivorous plants.

Cortex

In a plant, the cortex is a tissue of unspecialized cells lying between the epidermis (surface cells) and the vascular, or conducting, tissues of stems and roots. Cortical cells may contain stored carbohydrates or other substances such as resins, latex, essential oils, and tannins. In roots and in some herbaceous stems but not usually in woody stems, the innermost layer of cortical cells is differentiated into a cell layer called the endodermis. The cell walls of the endodermis possess a woody and corky band, called the casparian strip, around all the cell walls except those facing toward the axis and the surface of the root or stem. The endodermis with its casparian strips may function in regulating the flow of water between outer tissues and the vascular cylinder at the centre of the root. Within an inch or two of shoot tips, some flowering plants have a starch sheath (a layer of cells with much stored starch) in the same position as an endodermis.

The cortex often develops into a type of tissue called aerenchyma, which contains air spaces produced by separation, tearing, or dissolution of the cortex cell walls. Cortical cells in herbaceous stems, young woody stems, and stems of succulents (cacti and other fleshy plants) contain chloroplasts and can therefore convert carbon dioxide and water to simple carbohydrates (carbon fixation) using photosynthesis. Simple carbohydrates may then be metabolized into complex carbohydrates such as starch, which is stored in the cortex in edible roots, bulbs, and tubers.

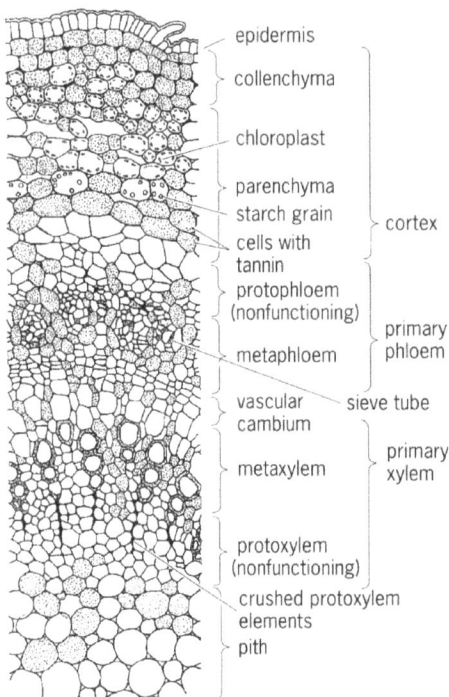

Transverse section of the Prunus stem showing the cortex which is composed of collenchyma and parenchyma.

Cortex is the mass of primary tissue in roots and stems extending inward from the epidermis to the phloem. The cortex may consist of one or a combination of three major tissues: parenchyma, collenchyma, and sclerenchyma. In roots the cortex almost always consists of parenchyma, and is bounded, more or less distinctly, by the hypodermis (exodermis) on the periphery and by the endodermis on the inside.

Cortical parenchyma is composed of loosely arranged thin-walled living cells. Prominent intercellular spaces usually occur in this tissue. In stems the cells of the outer parenchyma may appear green due to the presence of chloroplasts in the cells. This green tissue is sometimes called chlorenchyma, and it is probable that photosynthesis takes place in it.

In some species the cells of the outer cortex are modified in aerial stems by deposition of hemicellulose as an additional wall substance, especially in the corners or angles of the cells. This tissue is called collenchyma, and the thickening of the cell walls gives mechanical support to the shoot.

The cortex makes up a considerable proportion of the volume of the root, particularly in young roots, where it functions in the transport of water and ions from the epidermis to the vascular (xylem and phloem) tissues. In older roots it functions primarily as a storage tissue.

In addition to being supportive and protective, the cortex functions in the synthesis and localization of many chemical substances; it is one of the most fundamental storage tissues in the plant. The kinds of cortical cells specialized with regard to storage and synthesis are numerous.

Because the living protoplasts of the cortex are so highly specialized, patterns and gradients of many substances occur within the cortex, including starch, tannins, glucosides, organic acids, crystals of many kinds, and alkaloids. Oil cavities, resin ducts, and laticifers (latex ducts) are also common in the midcortex of many plants.

Plant Tissue Culture

Tissue culture is the In vitro aseptic culture of cells, tissues, organs or whole plant under controlled nutritional and environmental conditions often to produce the clones of plants. The resultant clones are true-to type of the selected genotype. The controlled conditions provide the culture an environment conducive for their growth and multiplication. These conditions include proper supply of nutrients, pH medium, adequate temperature and proper gaseous and liquid environment.

Plant tissue culture technology is being widely used for large scale plant multiplication. Apart from their use as a tool of research, plant tissue culture techniques have in recent years, become of major industrial importance in the area of plant propagation, disease elimination, plant improvement and production of secondary metabolites. Small pieces of tissue (named explants) can be used to produce hundreds and thousands of plants in a continuous process. A single explant can be multiplied into several thousand plants in relatively short time period and space under controlled conditions, irrespective of the season and weather on a year round basis. Endangered, threatened and rare species have successfully been grown and conserved by micropropagation because of high coefficient of multiplication and small demands on number of initial plants and space.

In addition, plant tissue culture is considered to be the most efficient technology for crop improvement by the production of somaclonal and gametoclonal variants. The micropropagation technology has a vast potential to produce plants of superior quality, isolation of useful variants in well-adapted high yielding genotypes with better disease resistance and stress tolerance capacities. Certain type of callus cultures give rise to clones that have inheritable characteristics different from those of parent plants due to the possibility of occurrence of somaclonal variability, which leads to the development of commercially important improved varieties. Commercial production of plants through micropropagation techniques has several advantages over the traditional methods of propagation through seed, cutting, grafting and air-layering etc. It is rapid propagation processes that can lead to the production of plants virus free. Coryodalisyanhusuo, an important medicinal plant was propagated by somatic embryogenesis from tuber-derived callus to produce disease free tubers. Meristem tip culture of banana plants devoid from banana bunchy top virus (BBTV) and brome mosaic virus (BMV) were produced. Higher yields have been obtained by culturing pathogen free germplasmIn vitro. Increase in yield up to 150% of virus-free potatoes was obtained in controlled conditions.

Basics of Plant Cell and Tissue Culture

In plant cell culture, plant tissues and organs are grown In vitro on artificial media, under aseptic and controlled environment. The technique depends mainly on the concept of totipotentiality of plant cells which refers to the ability of a single cell to express the full genome by cell division. Along with the totipotent potential of plant cell, the capacity of cells to alter their metabolism, growth and development is also equally important and crucial to regenerate the entire plant. Plant tissue culture medium contains all the nutrients required for the normal growth and development of plants. It is mainly composed of macronutrients, micronutrients, vitamins, other organic components, plant growth regulators, carbon source and some gelling agents in case of solid medium. Murashige and Skoog medium (MS medium) is most extensively used for the vegetative propagation of many plant species In vitro. The pH of the media is also important that affects both the growth of plants and activity of plant growth regulators. It is adjusted to the value between 5.4 - 5.8. Both the solid and liquid medium can be used for culturing. The composition of the medium, particularly the plant hormones and the nitrogen source has profound effects on the response of the initial explant.

Plant growth regulators (PGR's) play an essential role in determining the development pathway of plant cells and tissues in culture medium. The auxins, cytokinins and gibberellins are most commonly used plant growth regulators. The type and the concentration of hormones used depend mainly on the species of the plant, the tissue or organ cultured and the objective of the experiment. Auxins and cytokinins are most widely used plant growth regulators in plant tissue culture and their amount determined the type of culture established or regenerated. The high concentration of auxins generally favors root formation, whereas the high concentration of cytokinins promotes shoot regeneration. A balance of both auxin and cytokinin leads to the development of mass of undifferentiated cells known as callus.

Maximum root induction and proliferation was found in Stevia rebaudiana, when the medium is supplemented with 0.5 mg/l NAA. Cytokinins generally promote cell division and induce shoot formation and axillary shoot proliferation. High cytokinin to auxin ratio promotes shoot proliferation while high auxin to cytokinins ratio results in root formation. Shoot initiation and proliferation was found maximum, when the callus of black pepper was shifted to medium supplemented with BA at the concentration of 0.5 mg/l. Gibberellins are used for enhanced growth and to promote cell elongation. Maximum shoot length was observed in Phalaenopsis orchids when cultured in medium containing 0.5 mg/l GA_3.

Tissue Culture in Agriculture

As an emerging technology, the plant tissue culture has a great impact on both agriculture and industry, through providing plants needed to meet the ever increasing world demand. It has made significant contributions to the advancement of agricultural sciences in recent times and today they constitute an indispensable tool in modern agriculture.

Biotechnology has been introduced into agricultural practice at a rate without precedent. Tissue culture allows the production and propagation of genetically homogeneous, disease-free plant material. Cell and tissue In vitro culture is a useful tool for the induction of somaclonal variation. Genetic variability induced by tissue culture could be used as a source of variability to obtain new stable genotypes. Interventions of biotechnological approaches for In vitro regeneration, mass micropropagation techniques and gene transfer studies in tree species have been encouraging. In vitro cultures of mature and/or immature zygotic embryos are applied to recover plants obtained from inter-generic crosses that do not produce fertile seeds. Genetic engineering can make possible a number of improved crop varieties with high yield potential and resistance against pests. Genetic transformation technology relies on the technical aspects of plant tissue culture and molecular biology for:

- Production of improved crop varieties.

- Production of disease-free plants (virus).

- Genetic transformation.

- Production of secondary metabolites.

- Production of varieties tolerant to salinity, drought and heat stresses.

Germplasm Conservation

In vitro cell and organ culture offers an alternative source for the conservation of endangered genotypes. Germplasm conservation worldwide is increasingly becoming an essential activity due to the high rate of disappearance of plant species and the increased need for safeguarding the floristic patrimony of the countries. Tissue culture protocols can be used for preservation of vegetative tissues when the targets for conservation are clones instead of seeds, to keep the genetic background of a crop and to avoid the loss of the conserved patrimony due to natural disasters, whether biotic or abiotic stress. The plant species which do not produce seeds (sterile plants) or which have 're-calcitrant' seeds that cannot be stored for long period of time can successfully be preserved via In vitro techniques for the maintenance of gene banks.

Cryopreservation plays a vital role in the long-term In vitro conservation of essential biological material and genetic resources. It involves the storage of In vitro cells or tissues in liquid nitrogen that results incryo-injury on the exposure of tissues tophysical andchemical stresses. Successful cryopreservation is often ascertained by cell and tissue survival and the ability to re-grow or regenerate into complete plants or form new colonies. It is desirable to assess the genetic integrity of recovered germplasm to determine whether it is 'true-to-type' following cryopreservation. The fidelity of recovered plants can be assessed at phenotypic, histological, cytological, biochemical and molecular levels, although, there are advantages and limitations of the various approaches used to assess genetic stability. Cryobionomics is a new approach to study genetic stability in the cryopreserved plant materials. The embryonic tissues can be cryopreserved for future use or for germplasm conservation.

Embryo Culture

Embryo culture is a type of plant tissue culture that is used to grow embryos from seeds and ovules in a nutrient medium. In embryo culture, the plant develops directly from the embryo or indirectly through the formation of callus and then subsequent formation of shoots and roots. The technique has been developed to break seed dormancy, test the vitality of seeds, production of rare species and haploid plants. It is an effective technique that is employed to shorten the breeding cycle of plants by growing excised embryos and results in the reduction of long dormancy period of seeds. Intra-varietal hybrids of an economically important energy plant "Jatropha" have been produced successfully with the specific objective of mass multiplication. Somatic embryogenesis and plant regeneration has been carried out in embryo cultures of Jucara Palm for rapid cloning and improvement of selected individuals. In addition, conservation of endangered species can also be attained by practicing embryo culture technique. Recently a successful protocol has been developed for the In vitro propagation of Khayagrandifoliola by excising embryos from mature seeds. The plant has a high economic value for timber wood and for medicinal purposes as well. This technique has an important application in forestry by offering a mean of propagation of elite individuals where the selection and improvement of natural population is difficult.

Genetic Transformation

Genetic transformation is the most recent aspect of plant cell and tissue culture that provides the mean of transfer of genes with desirable trait into host plants and recovery of transgenic plants. The technique has a great potential of genetic improvement of various crop plants by integrating in plant biotechnology and breeding programmes. It has a promising role for the introduction of agronomically important traits such as increased yield, better quality and enhanced resistance to pests and diseases.

Genetic transformation in plants can be achieved by either vector-mediated (indirect gene transfer) or vectorless (direct gene transfer) method. Among vector dependant gene transfer methods, Agrobacterium-mediated genetic transformation is most widely used for the expression of foreign genes in plant cells. Successful introduction of agronomic traits in plants was achieved by using root explants for the genetic transformation. Virus-based vectors offers an alternative way of stable and rapid transient protein expression in plant cells thus providing an efficient mean of recombinant protein production on large scale.

Recently successful transgenic plants of Jatropha were obtained by direct DNA delivery to mature seed-derived shoot apices via particle bombardment method. This technology has an important impact on the reduction of toxic substances in seeds thus overcoming the obstacle of seed utilization in various industrial sector. Regeneration of disease or viral resistant plants is now achieved by employing genetic transformation technique. Researchers succeeded in developing transgenic plants of potato resistant to potato virus Y (PVY) which is a major threat to potato crop worldwide. In addition, marker free transgenic plants of Petunia hybrida were produced using multi-auto-transformation (MAT) vector system. The plants exhibited high level of resistance to Botrytis cinerea,causal agent of gray mold.

Protoplast Fusion

Somatic hybridization is an important tool of plant breeding and crop improvement by the production of interspecific and intergeneric hybrids. The technique involves the fusion of pro-toplasts of two different genomes followed by the selection of desired somatic hybrid cells and regeneration of hybrid plants. Protoplast fusion provides an efficient mean of gene transfer with desired trait from one species to another and has an increasing impact on crop improvement. So-matic hybrids were produced by fusion of protoplasts from rice and ditch reed using electrofusion treatment for salt tolerance.

In vitro fusion of protoplast opens a way of developing unique hybrid plants by overcoming the barriers of sexual incompatibility. The technique has been applicable in horticultural industry to create new hybrids with increased fruit yield and better resistance to diseases. Successful viable hybrid plants were obtained when protoplasts from citrus were fused with other related citrinae species. The potential of somatic hybridization in important crop plants is best illustrated by the production of intergeneric hybrid plants among the members of Brassicaceae. To resolvethe prob-lem of loss of chromosomes and decreased regeneration capacity, successful protocol has been established for the production of somatic hybrid plants by using two types of wheat protoplast as recipient and protoplast of Haynaldiavillosa as a fusion donor. It is also employed as an important gene source for wheat improvement.

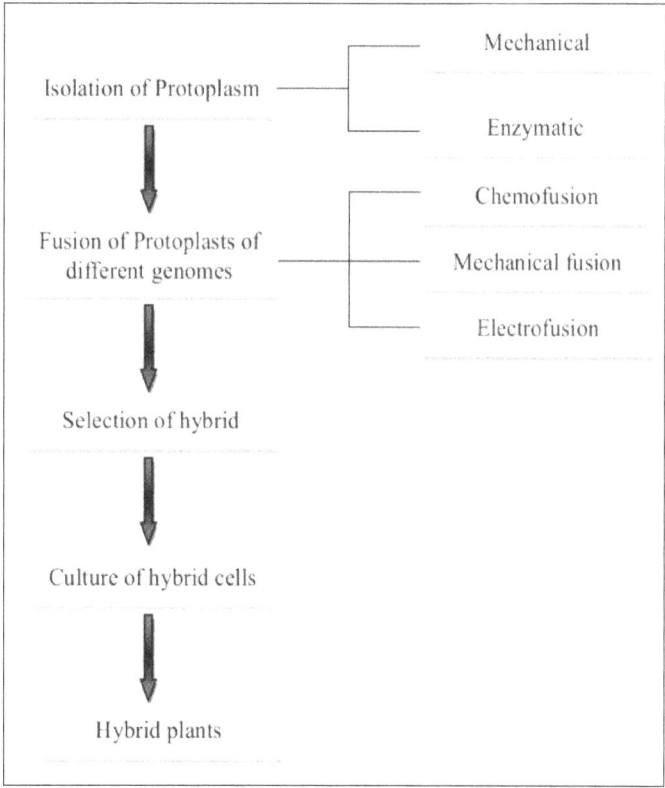

Schematic representation of production of hybrid plant via protoplast fusion.

Haploid Production

The tissue culture techniques enable to produce homozygous plants in relatively short time period through the protoplast, anther and microspore cultures instead of conventional breeding.

Haploids are sterile plants having single set of chromosomes which are converted into homozygous diploids by spontaneous or induced chromosome doubling. The doubling of chromosomes restores the fertility of plants resulting in production of double haploids with potential to become pure breeding new cultivars. The term androgenesis refers to the production of haploid plants from young pollen cells without undergoing fertilization. Sudherson et al. reported haploid plant production of sturt's desert pea by using pollen grains as primary explants. The haploidy technology has now become an integral part of plant breeding programs by speeding up the production of inbred lines and overcoming the constraints of seed dormancy and embryo non-viability. The technique has a remarkable use in genetic transformation by the production of haploid plants with induced resistance to various biotic and abiotic stresses. Introduction of genes with desired trait at haploid state followed by chromosome doubling led to the production of double haploids inbred wheat and drought tolerant plantswere attained successfully.

Techniques of Plant Tissue Culture

Micropropagation

Micropropagation starts with the selection of plant tissues (explant) from a healthy, vigorous mother plant. Any part of the plant (leaf, apical meristem, bud and root) can be used as explant.

Stage 0: Preparation of Donor Plant

Any plant tissue can be introduced In vitro. To enhance the probability of success, the mother plant should be ex vitro cultivated under optimal conditions to minimize contamination in the In vitro culture .

Stage I: Initiation Stage

In this stage an explant is surface sterilized and transferred into nutrient medium. Generally, the combined application of bactericide and fungicide products is suggested. The selection of products depends on the type of explant to be introduced. The surface sterilization of explant in chemical solutions is an important step to remove contaminants with minimal damage to plant cells. The most commonly used disinfectants are sodium hypochlorite, calcium hypochlorite, ethanol and mercuric chloride ($HgCl_2$). The cultures are incubated in growth chamber either under light or dark conditions according to the method of propagation.

Stage II: Multiplication Stage

The aim of this phase is to increase the number of propagules. The number of propagules is multiplied by repeated subcultures until the desired (or planned) number of plants is attained.

Stage III: Rooting Stage

The rooting stage may occur simultaneously in the same culture media used for multiplication of the explants. However, in some cases it is necessary to change media, including nutritional modification and growth regulator composition to induce rooting and the development of strong root growth.

Stage IV: Acclimatization Stage

At this stage, the In vitro plants are weaned and hardened. Hardening is done gradually from high to low humidity and from low light intensity to high light intensity. The plants are then transferred to an appropriate substrate (sand, peat, compost etc.) and gradually hardened under greenhouse.

Somatic Embryogenesis and Organogenesis

Somatic embryogenesis: is an In vitro method of plant regeneration widely used as an important biotechnological tool for sustained clonal propagation. It is a process by which somatic cells or tissues develop into differentiated embryos. These somatic embryos can develop into whole plants without undergoing the process of sexual fertilization as done by zygotic embryos. The somatic embryogenesis can be initiated directly from the explants or indirectly by the establishment of mass of unorganized cells named callus.

Plant regeneration via somatic embryogenesis occurs by the induction of embryogenic cultures from zygotic seed, leaf or stem segment and further multiplication of embryos.Mature embryos are then cultured for germination and plantlet development, and finally transferred to soil.

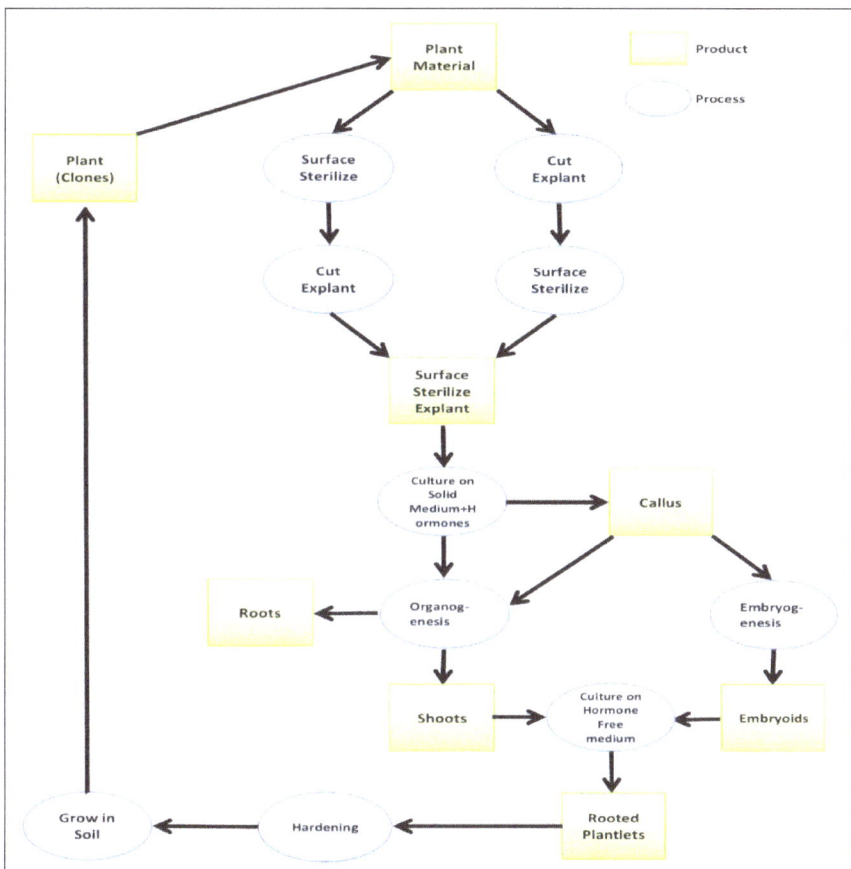

Flow chart summarizing tissue culture experiments.

Somatic embryogenesis has been reported in many plants including trees and ornamental plants of different families. The phenomenon has been observed in some cactus species. There are various factors that affect the induction and development of somatic embryos in cultured cells. A highly efficient protocol has been reported for somatic embryogenesis on grapevine that showed higher plant regeneration sufficiently when the tissues were cultured in liquid medium. Plant growth regulators play an important role in the regeneration and proliferation of somatic embryos. Highest efficiency of embryonic callus was induced by culturing nodal stem segments of rose hybrids on medium supplemented with various PGR's alone or in combination. This embryonic callus showed high germination rate of somatic embryos when grown on abscisic acid (ABA) alone. Somatic embryogenesis is not only a process of regenerating the plants for mass propagation but also regarded as a valuable tool for genetic manipulation. The process can also be used to develop the plants that are resistant to various kinds of stresses and to introduce the genes by genetic transformation. A successful protocol has been developed for regeneration of cotton cultivars with resistance to Fusarium and Verticillium wilts.

Organogenesis: refers to the production of plant organs i.e. roots, shoots and leaves that may arise directly from the meristem or indirectly from the undifferentiated cell masses (callus). Plant regeneration via organogenesis involves the callus production and differentiation of adventitious meristems into organs by altering the concentration of plant growth hormones in nutrient medium. Skoog and Muller were the first who demonstrated that high ratio of cytokinin to auxin stimulated the formation of shoots in tobacco callus while high auxin to cytokinin ratio induced root regeneration.

Tissue Culture in Pharmaceuticals

Plant cell and tissue cultures hold great promise for controlled production of myriad of useful secondary metabolites. Plant cell cultures combine the merits of whole-plant systems with those of microbial and animal cell cultures for the production of valuable therapeutic secondary metabolites. In the search for alternatives to production of medicinal compounds from plants, biotechnological approaches, specifically plant tissue cultures, are found to have potential as a supplement to traditional agriculture in the industrial production of bioactive plant metabolites. Exploration of the biosynthetic capabilities of various cell cultures has been carried out by a group of plant scientists and microbiologists in several countries during the last decade.

Cell suspension culture: Cell suspension culture systems are used now days for large scale culturing of plant cells from which secondary metabolites could be extracted. A suspension culture is developed by transferring the relatively friable portion of the callus into liquid medium and is maintained under suitable conditions of aeration, agitation, light, temperature and other physical parameters. Cell cultures cannot only yield defined standard phytochemicals in large volumes but also eliminate the presence of interfering compounds that occur in the field-grown plants. The advantage of this method is that it can ultimately provide a continuous, reliable source of natural products. The major advantage of the cell cultures include synthesis of bioactive secondary metabolites, running in controlled environment, independently from climate and soil conditions. A number of different types of bioreactors have been used for mass cultivation of plant cells. The first commercial application of large scale cultivation of plant cells was carried out in stirred tank reactors of 200 liter and 750 liter capacities to produce shikonin by cell culture of Lithospermumerythrorhizon. Cell of Catharanthusroseus, Dioscoreadeltoidea, Digitalis lanata, Panaxnotoginseng, Taxuswallichiana and Podophyllumhexandrum have been cultured in various bioreactors for the production of secondary plant products.

A number of medicinally important alkaloids, anticancer drugs, recombinant proteins and food additives are produced in various cultures of plant cell and tissues. Advances in the area of cell cultures for the production of medicinal compounds has made possible the production of a wide variety of pharmaceuticals like alkaloids, terpenoids, steroids, saponins, phenolics, flavanoids and amino acids. Some of these are now available commercially in the market for example shikonin and paclitaxel (Taxol). Until now 20 different recombinant proteins have been produced in plant cell culture, including antibodies, enzymes, edible vaccines, growth factors and cytokines . Advances in scale-up approaches and immobilization techniques contribute to a considerable increase in the number of applications of plant cell cultures for the production of compounds with a high added value. Some of the secondary plant products obtained from cell suspension culture of various plants are given in Table.

Table: List of some secondary plant product produced in suspension culture.

Secondary metabolite	Plant name
Vasine	Adhatodavasica
Artemisinin	Artemisia annua
Azadirachtin	Azadirachtaindica
Cathin	Bruceajavanica
Capsiacin	Capsicum annum
Sennosides	Cassia senna
Ajmalicine Secologanin Indole alkaloids Vincristine	Catharanthusroseus
Stilbenes	Cayratiatrifoliata
Berberin	Cosciniumfenustratum
Sterols	Hyssopusofficinalis
Shikonin	Lithospermumerythrorhizon
Ginseng saponin	Panaxnotoginseng
Podophyllotoxin	Podophyllumhexandrum
Taxane Paclitaxel	Taxuschinensis

Hairy Root Cultures

Table: List of some secondary plant product produced in Hairy root culture.

Secondary metabolite	Plant name
Rosmarinic acid	Agastacherugosa
Deoursin	Angelica gigas
Resveratol	Arachyshypogaea
Tropane	Brugmansia candida
Asiaticoside	Centellaasiatica
Rutin	Fagopyrumesculentum
Glucoside	Gentianamacrophylla
Glycyrrhizin	Glycyrrhizaglabra
Shikonin	Lithospermumerythrorhizon

Glycoside	Panax ginseng
Plumbagin	Plumbagozeylanica
Anthraquinone	Rubiaakane
Silymarin	Silybiummarianum
Flavonolignan	Silybiummariyanm
Vincamine	Vinca major
Withanoloid A	Withaniasomnifera

The hairy root system based on inoculation with Agrobacterium rhizogenes has become popular in the last two decades as a method of producing secondary metabolites synthesized in plant roots. Organized cultures, and especially root cultures, can make a significant contribution in the production of secondary metabolites. Most of the research efforts that use differentiated cultures instead of cell suspension cultures have focused on transformed (hairy) roots. Agrobacterium rhizogenes causes hairy root disease in plants. The neoplastic (cancerous) roots produced by A. rhizogenes infection are characterized by high growth rate, genetic stability and growth in hormone free media. High stability and productivity features allow the exploitation of hairy roots as valuable biotechnological tool for the production of plant secondary metabolites. These genetically transformed root cultures can produce levels of secondary metabolites comparable to that of intact plants. Hairy root technology has been strongly improved by increased knowledge of molecular mechanisms underlying their development. Optimizing the composition of nutrients for hairy root cultures is critical to gain a high production of secondary metabolites. Some of the secondary plant products obtained from hairy root culture of various plants are shown in table.

References

- Plant-tissue: biologydictionary.net, Retrieved 20 June, 2019
- Permanent-tissues-in-plants-essay-plant-anatomy-botany, plant-tissues-essay- 76998: biologydiscussion.com, Retrieved 15 July, 2019
- Dermal-tissue-epidermis, pharmacognosy-s-topics-plant-tissue: medicinalplants-pharmacognosy.com, Retrieved 15 May, 2019
- Cortex-plant-tissue: britannica.com, Retrieved 16 August, 2019
- Plant-tissue-culture-current-status-and-opportunities, recent-advances-in-plant-in-vitro-culture: intechopen.com, Retrieved 22 April, 2019
- Ewers, F.W. (1982). "Secondary growth in needle leaves of Pinus longaeva (bristlecone pine) and other conifers: Quantitative data". American Journal of Botany. 69: 1552–1559. doi:10.2307/2442909. JSTOR 2442909

3
Plant Cell Organelles

A plant cell is enclosed by a cell wall, containing a membrane-bound nucleus and other cell organelles. Protoplasm, cytoplasm, plastids, tannosome, etc. are some of these organelles. This chapter closely examines these plant cell organelles to provide an extensive understanding of the subject.

Cell Wall

Cell wall is specialized form of extracellular matrix that surrounds every cell of a plant. The cell wall is responsible for many of the characteristics that distinguish plant cells from animal cells. Although often perceived as an inactive product serving mainly mechanical and structural purposes, the cell wall actually has a multitude of functions upon which plant life depends. Such functions include: (1) providing the living cell with mechanical protection and a chemically buffered environment, (2) providing a porous medium for the circulation and distribution of water, minerals, and other small nutrient molecules, (3) providing rigid building blocks from which stable structures of higher order, such as leaves and stems, can be produced, and (4) providing a storage site of regulatory molecules that sense the presence of pathogenic microbes and control the development of tissues.

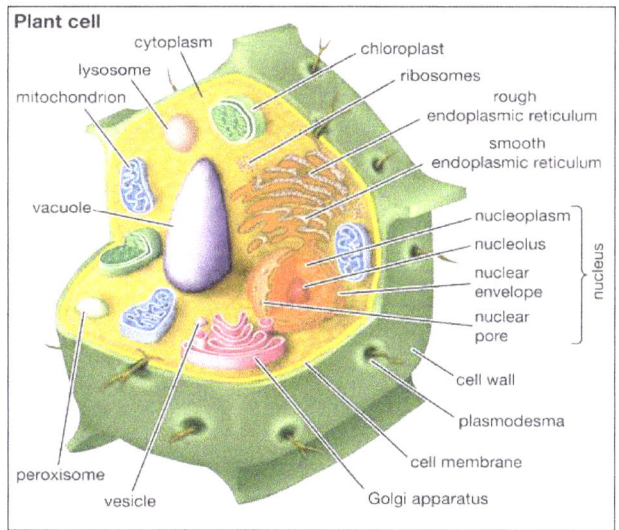

Cutaway drawing of a plant cell, showing the cell wall and internal organelles.

Certain prokaryotes, algae, slime molds, water molds, and fungi also have cell walls. Bacterial cell walls are characterized by the presence of peptidoglycan, whereas those of Archaea characteristically lack this chemical. Algal cell walls are similar to those of plants, and many contain specific polysaccharides that are useful for taxonomy. Unlike those of plants and algae, fungal cell walls lack cellulose entirely and contain chitin.

Mechanical Properties

All cell walls contain two layers, the middle lamella and the primary cell wall, and many cells produce an additional layer, called the secondary wall. The middle lamella serves as a cementing layer between the primary walls of adjacent cells. The primary wall is the cellulose-containing layer laid down by cells that are dividing and growing. To allow for cell wall expansion during growth, primary walls are thinner and less rigid than those of cells that have stopped growing. A fully grown plant cell may retain its primary cell wall (sometimes thickening it), or it may deposit an additional, rigidifying layer of different composition, which is the secondary cell wall. Secondary cell walls are responsible for most of the plant's mechanical support as well as the mechanical properties prized in wood. In contrast to the permanent stiffness and load-bearing capacity of thick secondary walls, the thin primary walls are capable of serving a structural, supportive role only when the vacuoles within the cell are filled with water to the point that they exert a turgor pressure against the cell wall. Turgor-induced stiffening of primary walls is analogous to the stiffening of the sides of a pneumatic tire by air pressure. The wilting of flowers and leaves is caused by a loss of turgor pressure, which results in turn from the loss of water from the plant cells.

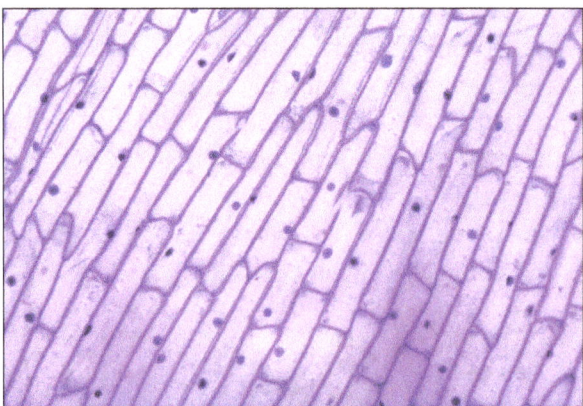

Plant cell Onion skin cells under a microscope.

Components

Although primary and secondary wall layers differ in detailed chemical composition and structural organization, their basic architecture is the same, consisting of cellulose fibres of great tensile strength embedded in a water-saturated matrix of polysaccharides and structural glycoproteins.

Cellulose

Cellulose is an organic compound with the formula $(C_{6H_{10}O_5})_n$, a polysaccharide consisting of a linear chain of several hundred to many thousands of $\beta(1\rightarrow4)$ linked D-glucose units. Cellulose is an important structural component of the primary cell wall of green plants, many forms of algae and the oomycetes. Some species of bacteria secrete it to form biofilms. Cellulose is the most abundant organic polymer on Earth. The cellulose content of cotton fiber is 90%, that of wood is 40–50%, and that of dried hemp is approximately 57%.

Cellulose is mainly used to produce paperboard and paper. Smaller quantities are converted into a wide variety of derivative products such as cellophane and rayon. Conversion of cellulose from energy crops into biofuels such as cellulosic ethanol is under development as a renewable fuel source. Cellulose for industrial use is mainly obtained from wood pulp and cotton.

Some animals, particularly ruminants and termites, can digest cellulose with the help of symbiotic micro-organisms that live in their guts, such as Trichonympha. In human nutrition, cellulose is a non-digestible constituent of insoluble dietary fiber, acting as a hydrophilic bulking agent for feces and potentially aiding in defecation.

Structure and Properties

Cellulose has no taste, is odorless, is hydrophilic with the contact angle of 20–30 degrees, is insoluble in water and most organic solvents, is chiral and is biodegradable. It was shown to melt at 467 °C in pulse tests made by Dauenhauer et al. (2016). It can be broken down chemically into its glucose units by treating it with concentrated mineral acids at high temperature.

Cellulose is derived from D-glucose units, which condense through $\beta(1{\to}4)$-glycosidic bonds. This linkage motif contrasts with that for $\alpha(1{\to}4)$-glycosidic bonds present in starch and glycogen. Cellulose is a straight chain polymer. Unlike starch, no coiling or branching occurs and the molecule adopts an extended and rather stiff rod-like conformation, aided by the equatorial conformation of the glucose residues. The multiple hydroxyl groups on the glucose from one chain form hydrogen bonds with oxygen atoms on the same or on a neighbor chain, holding the chains firmly together side-by-side and forming microfibrils with high tensile strength. This confers tensile strength in cell walls where cellulose microfibrils are meshed into a polysaccharide matrix. The high tensile strength of plant stems and of the tree wood also arises from the arrangement of cellulose fibers intimately distributed into the lignin matrix. The mechanical role of cellulose fibers in the wood matrix responsible for its strong structural resistance, can somewhat be compared to that of the reinforcement bars in concrete, lignin playing here the role of the hardened cement paste acting as the "glue" in between the cellulose fibers.

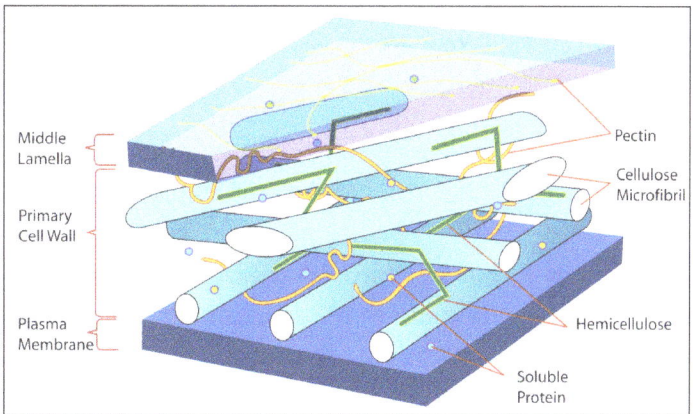

The arrangement of cellulose and other polysaccharides in a plant cell wall.

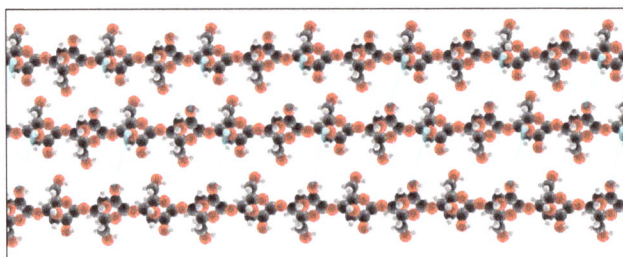

A triple strand of cellulose showing the hydrogen bonds (cyan lines) between glucose strands.

Cotton fibres represent the purest natural form of cellulose, containing more than 90% of this polysaccharide.

Compared to starch, cellulose is also much more crystalline. Whereas starch undergoes a crystalline to amorphous transition when heated beyond 60–70 °C in water (as in cooking), cellulose requires a temperature of 320 °C and pressure of 25 MPa to become amorphous in water.

Several different crystalline structures of cellulose are known, corresponding to the location of hydrogen bonds between and within strands. Natural cellulose is cellulose I, with structures I_α and I_β. Cellulose produced by bacteria and algae is enriched in I_α while cellulose of higher plants consists mainly of I_β. Cellulose in regenerated cellulose fibers is cellulose II. The conversion of cellulose I to cellulose II is irreversible, suggesting that cellulose I is metastable and cellulose II is stable. With various chemical treatments it is possible to produce the structures cellulose III and cellulose IV.

Many properties of cellulose depend on its chain length or degree of polymerization, the number of glucose units that make up one polymer molecule. Cellulose from wood pulp has typical chain lengths between 300 and 1700 units; cotton and other plant fibers as well as bacterial cellulose have chain lengths ranging from 800 to 10,000 units. Molecules with very small chain length resulting from the breakdown of cellulose are known as cellodextrins; in contrast to long-chain cellulose, cellodextrins are typically soluble in water and organic solvents.

Cellulose contains 44.44% carbon, 6.17% hydrogen, and 49.39% oxygen. The chemical formula of cellulose is $(C_6H_{10}O_5)n$ where n is the degree of polymerization and represents the number of glucose groups.

Plant-derived cellulose is usually found in a mixture with hemicellulose, lignin, pectin and other substances, while bacterial cellulose is quite pure, has a much higher water content and higher tensile strength due to higher chain lengths.

Cellulose is soluble in Schweizer's reagent, cupriethylenediamine (CED), cadmiumethylenediamine (Cadoxen), N-methylmorpholine N-oxide, and lithium chloride / dimethylacetamide. This is used in the production of regenerated celluloses (such as viscose and cellophane) from dissolving pulp. Cellulose is also soluble in many kinds of ionic liquids.

Cellulose consists of crystalline and amorphous regions. By treating it with strong acid, the amorphous regions can be broken up, thereby producing nanocrystalline cellulose, a novel material with many desirable properties. Recently, nanocrystalline cellulose was used as the filler phase in bio-based polymer matrices to produce nanocomposites with superior thermal and mechanical properties.

Processing

Assay

Given a cellulose-containing material, the carbohydrate portion that does not dissolve in a 17.5% solution of sodium hydroxide at 20 °C is α cellulose, which is true cellulose. Acidification of the extract precipitates β cellulose. The portion that dissolves in base but does not precipitate with acid is γ cellulose.

Cellulose can be assayed using a method described by Updegraff, where the fiber is dissolved in acetic and nitric acid to remove lignin, hemicellulose, and xylosans. The resulting cellulose is allowed to react with anthrone in sulfuric acid. The resulting coloured compound is assayed spectrophotometrically at a wavelength of approximately 635 nm.

In addition, cellulose can be represented by the difference between acid detergent fiber (ADF) and acid detergent lignin (ADL).

Luminescent conjugated oligothiophenes can also be used to detect cellulose using fluorescence microscopy or spectrofluorometric methods.

Biosynthesis

In plants cellulose is synthesized at the plasma membrane by rosette terminal complexes (RTCs). The RTCs are hexameric protein structures, approximately 25 nm in diameter, that contain the cellulose synthase enzymes that synthesise the individual cellulose chains. Each RTC floats in the cell's plasma membrane and "spins" a microfibril into the cell wall.

RTCs contain at least three different cellulose synthases, encoded by CesA (Ces is short for "cellulose synthase") genes, in an unknown stoichiometry. Separate sets of CesA genes are involved in primary and secondary cell wall biosynthesis. There are known to be about seven subfamilies in the plant CesA superfamily, some of which include the more cryptic, tentatively-named Csl (cellulose synthase-like) enzymes. These cellulose syntheses use UDP-glucose to form the $\beta(1\rightarrow4)$-linked cellulose.

Bacterial cellulose is produced using the same family of proteins, although the gene is called BcsA for "bacterial cellulose synthase" or CelA for "cellulose" in many instances. In fact, plants acquired CesA from the endosymbiosis event that produced the chloroplast. All cellulose synthases known belongs to glucosyltransferase family 2 (GT2).

Cellulose synthesis requires chain initiation and elongation, and the two processes are separate.

Cellulose synthase (CesA) initiates cellulose polymerization using a steroid primer, sitosterol-beta-glucoside, and UDP-glucose. It then utilizes UDP-D-glucose precursors to elongate the growing cellulose chain. A cellulase may function to cleave the primer from the mature chain.

Cellulose is also synthesised by tunicate animals, particularly in the tests of ascidians (where the cellulose was historically termed "tunicine" (tunicin)).

Breakdown

Cellulolysis is the process of breaking down cellulose into smaller polysaccharides called cellodextrins or completely into glucose units; this is a hydrolysis reaction. Because cellulose molecules bind strongly to each other, cellulolysis is relatively difficult compared to the breakdown of other polysaccharides. However, this process can be significantly intensified in a proper solvent, e.g. in an ionic liquid.

Most mammals have limited ability to digest dietary fiber such as cellulose. Some ruminants like cows and sheep contain certain symbiotic anaerobic bacteria (such as Cellulomonas and Ruminococcus spp.) in the flora of the rumen, and these bacteria produce enzymes called cellulases that help the microorganism to digest cellulose; the breakdown products are then used by the bacteria for proliferation. The bacterial mass is later digested by the ruminant in its digestive system (stomach and small intestine). Horses use cellulose in their diet by fermentation in their hindgut via symbiotic bacteria which produce cellulase to digest cellulose. Similarly, some termites contain in their hindguts certain flagellate protozoa producing such enzymes, whereas others contain bacteria or may produce cellulase.

The enzymes used to cleave the glycosidic linkage in cellulose are glycoside hydrolases including endo-acting cellulases and exo-acting glucosidases. Such enzymes are usually secreted as part of multienzyme complexes that may include dockerins and carbohydrate-binding modules.

At temperatures above 350 °C, cellulose undergoes thermolysis (also called 'pyrolysis'), decomposing into solid char, vapors, aerosols, and gases such as carbon dioxide. Maximum yield of vapors which condense to a liquid called bio-oil is obtained at 500 °C.

Semi-crystalline cellulose polymers react at pyrolysis temperatures (350–600 °C) in a few seconds; this transformation has been shown to occur via a solid-to-liquid-to-vapor transition, with the liquid (called intermediate liquid cellulose or molten cellulose) existing for only a fraction of a second. Glycosidic bond cleavage produces short cellulose chains of two-to-seven monomers comprising the melt. Vapor bubbling of intermediate liquid cellulose produces aerosols, which consist of short chain anhydro-oligomers derived from the melt.

Continuing decomposition of molten cellulose produces volatile compounds including levoglucosan, furans, pyrans, light oxygenates and gases via primary reactions. Within thick cellulose samples, volatile compounds such as levoglucosan undergo 'secondary reactions' to volatile products including pyrans and light oxygenates such as glycolaldehyde.

Hemicellulose

Hemicelluloses are polysaccharides related to cellulose that comprise about 20% of the biomass of land plants. In contrast to cellulose, hemicelluloses are derived from several sugars in addition to glucose, especially xylose but also including mannose, galactose, rhamnose, and arabinose. Hemicelluloses consist of shorter chains – between 500 and 3000 sugar units. Furthermore, hemicelluloses are branched, whereas cellulose is unbranched.

Derivatives

The hydroxyl groups (-OH) of cellulose can be partially or fully reacted with various reagents to afford derivatives with useful properties like mainly cellulose esters and cellulose ethers (-OR). In principle, though not always in current industrial practice, cellulosic polymers are renewable resources.

Ester derivatives include:

Cellulose ester	Reagent	Example	Reagent	Group R
Organic esters	Organic acids	Cellulose acetate	Acetic acid and acetic anhydride	H or $-(C=O)CH_3$
		Cellulose triacetate	Acetic acid and acetic anhydride	$-(C=O)CH_3$
		Cellulose propionate	Propionic acid	H or $-(C=O)CH_2CH_3$
		Cellulose acetate propionate (CAP)	Acetic acid and propanoic acid	H or $-(C=O)CH_3$ or $-(C=O)CH_2CH_3$
		Cellulose acetate butyrate (CAB)	Acetic acid and butyric acid	H or $-(C=O)CH_3$ or $-(C=O)CH_2CH_2CH_3$
Inorganic esters	Inorganic acids	Nitrocellulose (cellulose nitrate)	Nitric acid or another powerful nitrating agent	H or $-NO_2$
		Cellulose sulfate	Sulfuric acid or another powerful sulfuring agent	H or $-SO_3H$

The cellulose acetate and cellulose triacetate are film- and fiber-forming materials that find a variety of uses. The nitrocellulose was initially used as an explosive and was an early film forming material. With camphor, nitrocellulose gives celluloid.

Ether derivatives include:

Cellulose ethers	Reagent	Example	Reagent	Group R = H or	Water solubility	Application	E number
Alkyl	Halogenoalkanes	Methylcellulose	Chloromethane	$-CH_3$	Cold water-soluble		E461
		Ethylcellulose	Chloroethane	$-CH_2CH_3$	Water-insoluble	A commercial thermoplastic used in coatings, inks, binders, and controlled-release drug tablets	E462
		Ethyl methyl cellulose	Chloromethane and chloroethane	$-CH_3$ or $-CH_2CH_3$			E465

Hydroxy-alkyl	Epoxides	Hydroxyethyl cellulose	Ethylene oxide	-CH- $_2$CH$_2$OH	Cold/hot water-soluble	Gelling and thickening agent	
		Hydroxypropyl cellulose (HPC)	Propylene oxide	-CH- $_2$CH(OH) CH$_3$	Cold water-soluble		E463
		Hydroxyethyl methyl cellulose	Chloromethane and ethylene oxide	-CH$_3$ or -CH- $_2$CH$_2$OH	Cold water-soluble	Production of cellulose films	
		Hydroxypropyl methyl cellulose (HPMC)	Chloromethane and propylene oxide	-CH$_3$ or -CH- $_2$CH(OH) CH$_3$	Cold water-soluble	Viscosity modifier, gelling, foaming and binding agent	E464
		Ethyl hydroxyethyl cellulose	Chloroethane and ethylene oxide	-CH$_2$CH$_3$ or—CH- $_2$CH$_2$OH			E467
Carboxy-alkyl	Halogenated carboxylic acids	Carboxymethyl cellulose (CMC)	Chloroacetic acid	-CH- $_2$COOH	Cold/Hot water-soluble	Often used as its sodium salt, sodium carboxymethyl cellulose (NaCMC)	E466

The sodium carboxymethyl cellulose can be cross-linked to give the croscarmellose sodium (E468) for use as a disintegrant in pharmaceutical formulations.

Applications

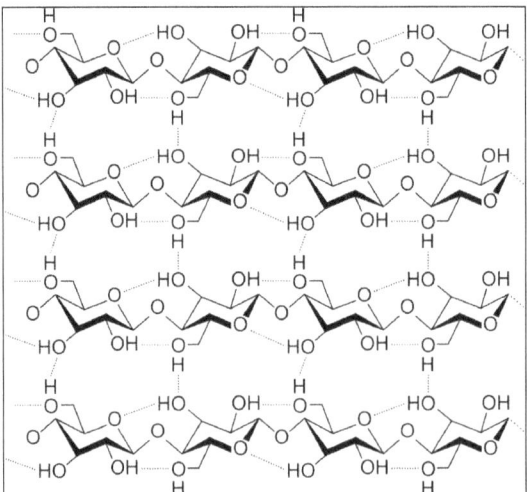

A strand of cellulose (conformation I$_\alpha$), showing the hydrogen bonds (dashed) within and between cellulose molecules.

Cellulose for industrial use is mainly obtained from wood pulp and cotton. The kraft process is used to separate cellulose from lignin, another major component of plant matter.

- Paper products: Cellulose is the major constituent of paper, paperboard, and card stock.

- Fibers: Cellulose is the main ingredient of textiles made from cotton, linen, and other plant fibers. It can be turned into rayon, an important fiber that has been used for textiles since the beginning of the 20th century. Both cellophane and rayon are known as "regenerated cellulose fibers"; they are identical to cellulose in chemical structure and are usually made from dissolving pulp via viscose. A more recent and environmentally friendly method to produce a form of rayon is the Lyocell process.

- Consumables: Microcrystalline cellulose (E460i) and powdered cellulose (E460ii) are used as inactive fillers in drug tablets and a wide range of soluble cellulose derivatives, E numbers E461 to E469, are used as emulsifiers, thickeners and stabilizers in processed foods. Cellulose powder is, for example, used in processed cheese to prevent caking inside the package. Cellulose occurs naturally in some foods and is an additive in manufactured foods, contributing an indigestible component used for texture and bulk, potentially aiding in defecation.

- Science: Cellulose is used in the laboratory as a stationary phase for thin layer chromatography. Cellulose fibers are also used in liquid filtration, sometimes in combination with diatomaceous earth or other filtration media, to create a filter bed of inert material.

- Energy crops: The major combustible component of non-food energy crops is cellulose, with lignin second. Non-food energy crops produce more usable energy than edible energy crops (which have a large starch component), but still compete with food crops for agricultural land and water resources. Typical non-food energy crops include industrial hemp (though outlawed in some countries), switchgrass, Miscanthus, Salix (willow), and Populus (poplar) species.

- Biofuel: TU-103, a strain of Clostridium bacteria found in zebra waste, can convert nearly any form of cellulose into butanol fuel.

- Building material: Hydroxyl bonding of cellulose in water produces a sprayable, moldable material as an alternative to the use of plastics and resins. The recyclable material can be made water- and fire-resistant. It provides sufficient strength for use as a building material. Cellulose insulation made from recycled paper is becoming popular as an environmentally preferable material for building insulation. It can be treated with boric acid as a fire retardant.

- Miscellaneous: Cellulose can be converted into cellophane, a thin transparent film. It is the base material for the celluloid that was used for photographic and movie films until the mid-1930s. Cellulose is used to make water-soluble adhesives and binders such as methyl cellulose and carboxymethyl cellulose which are used in wallpaper paste. Cellulose is further used to make hydrophilic and highly absorbent sponges. Cellulose is the raw material in the manufacture of nitrocellulose (cellulose nitrate) which is used in smokeless gunpowder.

- Pharmaceuticals: Cellulose derivatives, such as microcrystalline cellulose (MCC), have the advantages of retaining water, being a stabilizer and thickening agent, and in reinforcement of drug tablets.

Matrix Polysaccharides

The two major classes of cell wall matrix polysaccharides are the hemicelluloses and the pectic polysaccharides, or pectins. Both are synthesized in the Golgi apparatus, brought to the cell surface in small vesicles, and secreted into the cell wall.

Hemicelluloses consist of glucose molecules arranged end to end as in cellulose, with short side chains of xylose and other uncharged sugars attached to one side of the ribbon. The other side of the ribbon binds tightly to the surface of cellulose fibrils, thereby coating the microfibrils with hemicellulose and preventing them from adhering together in an uncontrolled manner. Hemicellulose molecules have been shown to regulate the rate at which primary cell walls expand during growth.

The heterogeneous, branched, and highly hydrated pectic polysaccharides differ from hemicelluloses in important respects. Most notably, they are negatively charged because of galacturonic acid residues, which, together with rhamnose sugar molecules, form the linear backbone of all pectic polysaccharides. The backbone contains stretches of pure galacturonic acid residues interrupted by segments in which galacturonic acid and rhamnose residues alternate; attached to these latter segments are complex, branched sugar side chains. Because of their negative charge, pectic polysaccharides bind tightly to positively charged ions, or cations. In cell walls, calcium ions cross-link the stretches of pure galacturonic acid residues tightly, while leaving the rhamnose-containing segments in a more open, porous configuration. This cross-linking creates the semirigid gel properties characteristic of the cell wall matrix—a process exploited in the preparation of jellied preserves.

Proteins

Although plant cell walls contain only small amounts of protein, they serve a number of important functions. The most prominent group are the hydroxyproline-rich glycoproteins, shaped like rods with connector sites, of which extensin is a prominent example. Extensin contains 45 percent hydroxyproline and 14 percent serine residues distributed along its length. Every hydroxyproline residue carries a short side chain of arabinose sugars, and most serine residues carry a galactose sugar. This gives rise to long molecules, resembling bottle brushes, that are secreted into the cell wall toward the end of primary wall formation and become covalently cross-linked into a mesh at the time that cell growth stops. Plant cells may control their ultimate size by regulating the time at which this cross-linking of extensin molecules occurs.

In addition to the structural proteins, cell walls contain a variety of enzymes. Most notable are those that cross-link extensin, lignin, cutin, and suberin molecules into networks. Other enzymes help protect plants against fungal pathogens by breaking fragments off of the cell walls of the fungi. The fragments in turn induce defense responses in underlying cells. The softening of ripe fruit and dropping of leaves in the autumn are brought about by cell wall-degrading enzymes.

Plastics

Cell wall plastics such as lignin, cutin, and suberin all contain a variety of organic compounds cross-linked into tight three-dimensional networks that strengthen cell walls and make them more resistant to fungal and bacterial attack. Lignin is the general name for a diverse group of polymers of aromatic alcohols. Deposited mostly in secondary cell walls and providing the rigidity of terrestrial vascular plants, it accounts for up to 30 percent of a plant's dry weight. The diversity of cross-links between the polymers—and the resulting tightness—makes lignin a formidable barrier to the penetration of most microbes.

Cutin and suberin are complex biopolyesters composed of fatty acids and aromatic compounds. Cutin is the major component of the cuticle, the waxy, water-repelling surface layer of cell walls exposed to the environment aboveground. By reducing the wettability of leaves and stems—and thereby affecting the ability of fungal spores to germinate—it plays an important part in the defense strategy of plants. Suberin serves with waxes as a surface barrier of underground parts. Its synthesis is also stimulated in cells close to wounds, thereby sealing off the wound surfaces and protecting underlying cells from dehydration.

Intercellular Communication

Plasmodesmata

Similar to the gap junction of animal cells is the plasmodesma, a channel passing through the cell wall and allowing direct molecular communication between adjacent plant cells. Plasmodesmata are lined with cell membrane, in effect uniting all connected cells with one continuous cell membrane. Running down the middle of each channel is a thin membranous tube that connects the endoplasmic reticula (ER) of the two cells. This structure is a remnant of the ER of the original parent cell, which, as the parent cell divided, was caught in the developing cell plate.

Although the precise mechanisms are not fully understood, the plasmodesma is thought to regulate the passage of small molecules such as salts, sugars, and amino acids by constricting or dilating the openings at each end of the channel.

Oligosaccharides with Regulatory Functions

The discovery of cell wall fragments with regulatory functions opened a new era in plant research. For years scientists had been puzzled by the chemical complexity of cell wall polysaccharides, which far exceeds the structural requirements of plant cell walls. The answer came when it was found that specific fragments of cell wall polysaccharides, called oligosaccharins, are able to induce specific responses in plant cells and tissues. One such fragment, released by enzymes used by fungi to break down plant cell walls, consists of a linear polymer of 10 to 12 galacturonic acid residues. Exposure of plant cells to such fragments induces them to produce antibiotics known as phytoalexins. In other experiments it has been shown that exposing strips of tobacco stem cells to a different type of cell wall fragment leads to the growth of roots; other fragments lead to the formation of stems and yet others to the production of flowers. In all instances the concentration of oligosaccharins required to bring about the observed responses is equal to that of hormones in animal cells; indeed, oligosaccharins may be viewed as the oligosaccharide hormones of plants.

Lignin

Lignin is one of the main components of plant cell wall and it is a natural phenolic polymer with high molecular weight, complex composition and structure.

Lignin is one of the most important secondary metabolite which is produced by the phenylalanine/tyrosine metabolic pathway in plant cells. It is the second most profuse biopolymers that accounts for 30% of the organic carbon content in biosphere. Lignin biosynthesis is a very complex network that is divided into three processes: (i) biosynthesis of lignin monomers, (ii) transport and (iii) polymerization. After a series of steps involving deamination, hydroxylation, methylation and reduction, lignin monomers are produced in cytoplasm and transported to the apoplast. Finally, lignin is generally polymerized with three main types of monolignols (sinapyl alcohol, S unit; coniferyl alcohol, G unit and p-coumaryl alcohol, H unit) by peroxidase (POD) and laccase (LAC) in secondary cell wall. In addition, several other compounds including hydroxycinnamaldehydes, tricin flavones, hydroxystilbenes and xenobiotics etc. have also been recognized to be lignin subunits. In differentiating protoxylem tracheary elements of Arabidopsis, lignin monomers can be free to diffuse in the extracellular space but are only polymerized in the secondary cell walls.

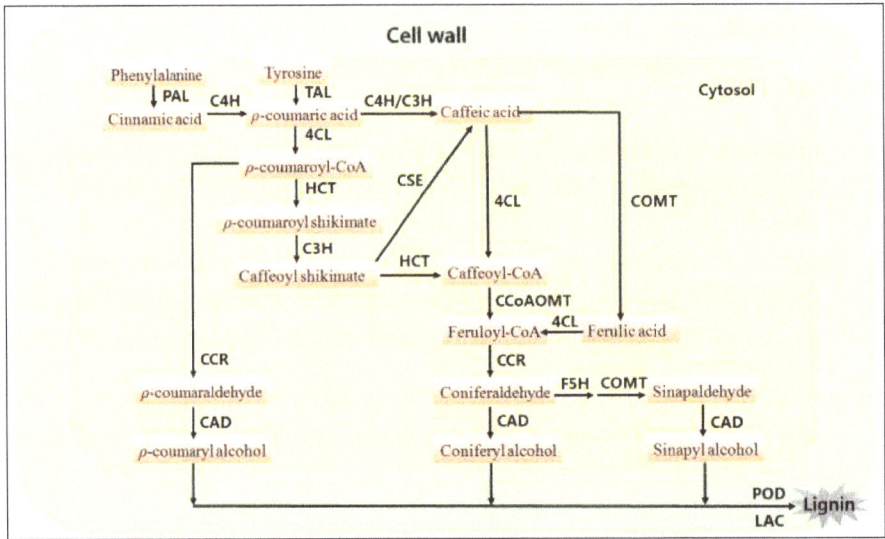

The general biosynthesis pathway of lignin in higher plants. PAL, phenylalanine ammonia-lyase; TAL, tyrosine ammonia-lyase; C4H, cinnamate 4-hydroxylase; 4CL, 4-coumarate: CoA ligase; CCR, cinnamoyl-CoA reductase; HCT, hydroxycinnamoyl-CoA shikimate/ Quinatehydroxycinnamoyltransferase; C3H, p-coumarate 3-hydroxylase; CCoAOMT, caffeoyl-CoA O-methyltransferase; F5H, ferulate 5-hydroxylase; CSE, caffeoyl shikimate esterase; COMT, caffeic acid O-methyltransferase; CAD, cinnamyl alcohol dehydrogenase; LAC, laccase; POD, peroxidase.

Lignin and its related metabolism play important roles in the growth and development of plants. As a complex phenolic polymer, lignin enhances plant cell wall rigidity, hydrophobic properties and promotes minerals transport through the vascular bundles in plant. In addition, lignin is an important barrier that protects against pests and pathogens. Lignin metabolism can also be actively involved in plant lodging resistance and in response to various environmental stresses.

As one of the important components of plant cell wall, lignin is of great significance to plant growth and environmental adaptability. In addition, lignin itself can also be used as a resource for the field of energy or pharmaceutical industry.

Genetic Modification of Lignin Biosynthesis

Lignin content, composition and structure can be altered using genetic modification. The effects of single or multiple lignin biosynthesis gene expression on phenotypic traits of transgenic plants can be directly observed and identified. In the past two decades, many researchers have used lignin engineering to modify lignin content and composition in plants. Lignin content in Arabidopsis thaliana quadruple mutant (pal1/pal2/pal3/pal4) of PAL genes, decreased by 20–25% as compared with the wild type, while the quadruple mutant also displayed reduced levels of salicylic acid and increased susceptibility to pathogen. Gui et al. found that the inhibition of Os4CL3 gene expression significantly reduced the lignin content and height of rice plants. Suppression of the mRNA level of 4CL gene in Pinus radiata reduced the lignin content of transgenic plants by 36–50%, which mainly due to depletion of guaiacyl lignin and resulted in a dwarfed phenotype of plant. Cinnamoyl-CoA reductase (CCR) and cinnamyl alcohol dehydrogenase (CAD) are the last two enzymes that function in the monolignols synthesis pathway, while the disruption of their function also affects the lignin content as found in the Arabidopsis thaliana ccc triple mutant (cad c cad d ccr1), i.e., lignin content was reduced by 50% as compared with the wild type and was accompanied by the male-sterile phenotype. Recently, Van Acker et al. found that CAD1-deficiency resulted in different metabolic routes for coniferaldehyde and sinapaldehyde and modified lignin content and structure in poplar. In C4 forage grasses, lignin deposition in the thick-walled parenchyma bundle-sheath cells affects digestibility of forage by animals, down-regulation of CCR gene is considered to be an effective strategy for the production of low-lignin C4 Paspalumdilatatum. Hydroxycinnamoyl: CoA transferase (HCT), p-coumarate 3-hydroxylase (C3H) and caffeate/5-hydroxyconiferaldehyde O-methyltransferase (COMT) are involved in the synthesis of sinapyl alcohol (S unit) and coniferyl alcohol (G unit).

Interference of the expression of CCR1 and COMT1 genes significantly altered lignin content and composition in ryegrass and enhanced digestibility without significant negative effects on either plant fitness or biomass production. Shadle et al. showed that transgenic Medicago sativa, expressing an HCT antisense construct, led to a significant decrease in lignin content and an obvious change in the lignin composition, while exhibited obvious stunting, decreased biomass and delayed flowering. Vanholme et al. found that Populus nigra hct1 mutant had a modified H lignin disposition. Inhibition of C3H expression in poplars also reduced the lignin content of poplar and changed its S/G ratio. Zhang et al. reported that overexpression of the monolignol 4-O-methyltransferase MOMT4 gene resulted in a 24% reduction in lignin content in Arabidopsis cells and increased saccharification yields of transgenic plants. Ferulate 5-hydroxylase (F5H) is one of key enzymes that regulate S/G lignin composition in plants, recent study has shown that the heterologous expression of angiosperm F5H induced ectopic productions of S lignin in the gymnosperm cell walls. In rice, overexpression of a F5H gene OsCAld5H1 increased the content of S units, while down-regulation of it enhanced the production of G lignin. Recently, caffeoyl shikimate esterase (CSE) was reported to be involved in lignin biosynthesis in Arabidopsis thaliana, Medicago truncatula and hybrid poplar. In addition, tricin was shown to react with monolignols, and down-regulated chalcone synthase (CHS) significantly reduced the contents of apigenin- and tricin-related flavonoids, resulting in a strongly reduced tricin lignin. Eudes et al. found that Sorghum bicolor (SbCOMT) can methylate the tricin precursors (luteolin and selgin) and be involved in the biosynthesis of lignin-linked tricin and S lignin units in sorghum.

After the synthesis of monomers in the cytoplasm, they are transported across the cell membrane and then polymerized in the cell wall. Arabidopsis AtABCG29 knockout mutants exhibited less lignin content and more sensitivity to p-coumaryl alcohol, due to p-coumaryl alcohol transport activity of AtABCG29 protein. Peroxidases and laccases are two key enzymes that participate in the polymerization of monomers. Overexpression of POD increases the content of phenol and lignin in plants. Shigeto et al. found that AtPrx double mutants (atprx2/atprx25, atprx2/atprx71 and atprx25/atprx71) had lower lignin content than single mutants but did not seriously affect the growth. Recently, a new class III peroxidase PRX17 (At2g22420) in Arabidopsis thaliana was reported to participate in lignin biosynthesis and be regulated by a MADS-box transcription factor AGL15. Compared to above mentioned lignin biosynthesis genes, the role of laccase in lignin biosynthesis is relatively poorly studied in the past years. Wang et al. confirmed that the overexpression of cotton laccase gene significantly increased the total lignin content of transgenic poplar. Berthet et al. found that lignin content in the stems of two Arabidopsis laccase double mutants, lac4-1 lac17 and lac4-2 lac17, decreased by 20% and 40%, respectively. Surprisingly, lignin deposition in roots has almost completely disappeared in the Arabidopsis lac11 lac4 lac17 triple mutant. Recently, a new lignification-related LACCASE5 was identified in Brachypodium distachyon, which lead to a modification on lignin content and composition.

Studies have shown that phenylpropanoid metabolism and lignin-specific synthesis can be regulated by transcription factors. The most widely studied transcription factor for lignin biosynthesis is the MYB (v-myb avian myeloblastosis viral oncogene homolog) family. It has been found that overexpression of Arabidopsis thaliana MYB58 and MYB63 transcription factors activated the expression of lignin biosynthesis-related genes and promoted the lignification of cells. While the other transcription factors, such as EgMYB1 from eucalyptus, MusaMYB31 from Musa inhibited the expression of genes related to lignin biosynthesis and negatively regulated lignin accumulation. In addition to the MYB transcription factor, some members of NAC transcription factor family in Arabidopsis were able to control lignin biosynthesis by regulating the entire cell wall synthesis-related genes. WRKY transcription factors can also work as regulators of lignin biosynthesis genes. Down-regulation of a WRKY transcription factor exhibited increased lignin level and enhanced biomass yield in Medicago sativa L. In the last few years, more and more lignin metabolism related genes have been identified by in vivo experiments, such as maize Caffeoyl-CoA O-methyl transferase gene CCoAOMT1, switchgrass UDP-Arabinomutase gene, Betula platyphylla MADS-box gene BpMADS12 and Eriobotrya japonica heat shock factors EjHSF3.

Role of Lignin in Plant Growth and Development

As one of the main components of the plant cell wall, lignin confers to the function of multiple types of cells in plants tissues and organs. Lignin metabolism is involved in plant growth and development, interference of lignin biosynthesis, especially H units, often leading to inhibition of plant growth and deformity development. In some plants, lignin accumulation is important for the seed propagation. The deposition of lignin in seed coat can protect the seeds from external adverse factors. Liang et al. found that the lower content of lignin in Arabidopsis thaliana mutant seed coat significantly decreased the seed germination rate, in comparison to wild type. The content of lignin in the Arabidopsis thaliana CCR1 mutant was significantly reduced, accompanied by stunted growth and reduced number of seeds. In Arabidopsis thalianaC4H mutant, the growth and lignin accumulation was inhibited, apical dominance was lost and showed male sterility. Similarly, simultaneous disruption of CAD and CCR leading to an obvious decrease in the content of lignin in Arabidopsis thaliana and a change in lignin composition, resulting in severe suppression of plant growth and male sterility, which may be related to lack of lignin in the anther. Herrero et al. found that in comparison with the wild type, total lignin and S unit contents were obviously decreased in

Arabidopsis thaliana peroxidase AtPrx72 knockout mutant, which had slow growth, less branches, smaller flower and stem than the wild-type plants, as well as significantly reduced photosynthetic efficiency. However, the mechanisms of lignin affecting plant growth and development are poorly understood. Recently, Bonawitz et al. have suggested that the transcriptional process and signaling pathways responding to cell wall defects may play an important role in lignin-deficient induced stunted growth.

Role of Lignin in Plant Lodging Resistance

Lodging resistance can prevent plant stems from bending or breaking, it is one of the most important traits that affect crop growth and grain yield. Numerous studies have shown that the lodging resistance of crops is related to plant height, biomass, stem diameter and the composition and characteristics of stem cell walls. Lignin accumulation in cell wall significantly enhance the mechanical strength of plant stalks. It has important implications for crop lodging resistance. Peng et al. added exogenous paclobutrazol to lodging-resistant/susceptible cultivar of winter wheat and found that paclobutrazol significantly reduced the internode length of wheat, promoted the lateral growth and increased lignin deposition, the activities of lignin biosynthesis enzymes and thickness of internode, thus improved the wheat lodging tolerance. Another study reported that the crop density can significantly change the morphological characteristics and the lignin biosynthesis of the stem and thus enhance the mechanical strength of the stem and reduce the hazard of lodging. Hu et al. reported that the lignin content and lignin biosynthesis enzymes (PAL, 4CL, CAD and POD) activities had important roles in the lodging resistance according to the analysis of lignin metabolism related indexes in Fagopyrum esculentum Moench varieties with a different lodging resistance.

In addition, nutrient elements have an important effect on plant lignin biosynthesis and lodging resistance. Silicon can enhance the expression of rice CAD gene, improve the accumulation of lignin and increase the strength of the stalk, thereby enhancing the lodging resistance. However, excessive nitrogen fertilizer significantly decreased the mechanical strength of stalk and the lodging resistance by reducing lignin biosynthesis in buckwheat, rapeseed and japonica rice. Kong et al. found that the addition of K^+ could significantly alleviate the effect of high NH_4^+ on the wheat culm strength. Therefore, it was suggested that the decrease of lodging resistance induced by nitrogen fertilizer was possibly related to the inhibition of uptake of K^+ which increased the lignin accumulation in the vascular bundles.

Different Types of Lignin and its Application

As a cheap, renewable resource, plant lignin is mostly used as an energy substance, or to develop new materials, e.g., lignin-based carbon fibers, due to the presence of phenolic hydroxyl groups and aliphatic hydroxyl groups in lignin structures. Lignin can be generally divided into three types according to the different plant species: softwood, hardwood and grass lignin. Softwood lignin consists exclusively of coniferyl alcohol, hardwood lignin consists mainly of coniferyl alcohol and sinapyl alcohol, grass lignin has three types of monomers (coniferyl, sinapyl and p-coumaryl alcohol). In addition, the non-conventional types such as the caffeyl lignin (C-lignin) were found in the seeds of vanilla orchid and several species of the Cactaceae.

The composition of the monomers has an important influence on the molecular structure of lignin, such as branching of the polymer and the degree of crosslinking with the polysaccharide.

Therefore, the monomer composition determines the degradability of lignin and the workability of lignocellulosic biomass. For example, corn and flax lignin generally contain high content of aliphatic OH groups and non-methoxylate phenolic groups respectively, which are suitable for production of phenolic resins and polyurethane synthesis, respectively; the content of OH groups is balanced in triticale lignin, which is appropriate for polyester synthesis. In addition, low molecular weight monomers derived from H and G units have a certain antioxidant capacity, making lignin also has some biological activity, such as anti-tumor.

Pectin

Pectin is a structural acidic heteropolysaccharide contained in the primary cell walls of terrestrial plants. Its main component is galacturonic acid, a sugar acid derived from galactose. It was first isolated and described in 1825 by Henri Braconnot. It is produced commercially as a white to light brown powder, mainly extracted from citrus fruits, and is used in food as a gelling agent, particularly in jams and jellies. It is also used in dessert fillings, medicines, sweets, as a stabilizer in fruit juices and milk drinks, and as a source of dietary fiber.

Commercially produced powder of pectin, extracted from citrus fruits.

Biology

In plant biology, pectin consists of a complex set of polysaccharides that are present in most primary cell walls and are particularly abundant in the non-woody parts of terrestrial plants. Pectin is a major component of the middle lamella, where it helps to bind cells together, but is also found in primary cell walls. Pectin is deposited by exocytosis into the cell wall via vesicles produced in the golgi.

The amount, structure and chemical composition of pectin differs among plants, within a plant over time, and in various parts of a plant. Pectin is an important cell wall polysaccharide that allows primary cell wall extension and plant growth. During fruit ripening, pectin is broken down by the enzymes pectinase and pectinesterase, in which process the fruit becomes softer as the middle lamellae break down and cells become separated from each other. A similar process of cell separation caused by the breakdown of pectin occurs in the abscission zone of the petioles of deciduous plants at leaf fall.

Pectin is a natural part of the human diet, but does not contribute significantly to nutrition. The daily intake of pectin from fruits and vegetables can be estimated to be around 5 g if approximately

500 g of fruits and vegetables are consumed per day.

In human digestion, pectin binds to cholesterol in the gastrointestinal tract and slows glucose absorption by trapping carbohydrates. Pectin is thus a soluble dietary fiber. In non-obese diabetic (NOD) mice pectin has been shown to increase the incidence of diabetes.

A study found that after consumption of fruit the concentration of methanol in the human body increased by as much as an order of magnitude due to the degradation of natural pectin which is esterified with methyl alcohol in the colon.

Pectin has been observed to have some function in repair the DNA of some types of plant seeds, usually desert plants. Pectinaceous surface pellicles, which are rich in pectin, create a mucilage layer that holds in dew that helps the cell repair its DNA.

Consumption of pectin has been shown to slightly (3-7%) reduce blood LDL cholesterol levels. The effect depends upon the source of pectin; apple and citrus pectins were more effective than orange pulp fiber pectin. The mechanism appears to be an increase of viscosity in the intestinal tract, leading to a reduced absorption of cholesterol from bile or food. In the large intestine and colon, microorganisms degrade pectin and liberate short-chain fatty acids that have positive influence on health (prebiotic effect).

Chemistry

Pectins, also known as pectic polysaccharides, are rich in galacturonic acid. Several distinct polysaccharides have been identified and characterised within the pectic group. Heterogalacturonans are linear chains of α-(1–4)-linked D-galacturonic acid. Substituted galacturonans are characterized by the presence of saccharide appendant residues (such as D-xylose or D-apiose in the respective cases of xylogalacturonan and apiogalacturonan) branching from a backbone of D-galacturonic acid residues. Rhamnogalacturonan I pectins (RG-I) contain a backbone of the repeating disaccharide: 4)-α-D-galacturonic acid-(1,2)-α-L-rhamnose-(1. From many of the rhamnose residues, sidechains of various neutral sugars branch off. The neutral sugars are mainly D-galactose, L-arabinose and D-xylose, with the types and proportions of neutral sugars varying with the origin of pectin.

Another structural type of pectin is rhamnogalacturonan II (RG-II), which is a less frequent, complex, highly branched polysaccharide. Rhamnogalacturonan II is classified by some authors within the group of substituted galacturonans since the rhamnogalacturonan II backbone is made exclusively of D-galacturonic acid units.

Isolated pectin has a molecular weight of typically 60,000–130,000 g/mol, varying with origin and extraction conditions.

In nature, around 80 percent of carboxyl groups of galacturonic acid are esterified with methanol. This proportion is decreased to a varying degree during pectin extraction. Pectins are classified as high- vs. low-methoxy pectins (short HM-pectins vs. LM-pectins), with more or less than half of all the galacturonic acid esterified. The ratio of esterified to non-esterified galacturonic acid determines the behavior of pectin in food applications - HM-pectins can form a gel under acidic conditions in the presence of high sugar concentrations, while LM-pectins form gels by interaction with divalent cations, particularly Ca^{2+}, according to the idealized 'egg box' model, in which ionic bridges are formed between calcium ions and the ionised carboxyl groups of the galacturonic acid.

In high-ester/high-methoxy pectins at soluble solids content above 60% and a pH-value between 2.8 and 3.6, hydrogen bonds and hydrophobic interactions bind the individual pectin chains together. These bonds form as water is bound by sugar and forces pectin strands to stick together. These form a 3-dimensional molecular net that creates the macromolecular gel. The gelling-mechanism is called a low-water-activity gel or sugar-acid-pectin gel.

While low-ester/low-methoxy pectins need calcium to form a gel, they can do so at lower soluble solids and higher pH-values than high-ester pectins. Normally low-ester pectins form gels with a range of pH from 2.6 to 7.0 and with a soluble solids content between 10 and 70%.

The non-esterified galacturonic acid units can be either free acids (carboxyl groups) or salts with sodium, potassium, or calcium. The salts of partially esterified pectins are called pectinates, if the degree of esterification is below 5 percent the salts are called pectates, the insoluble acid form, pectic acid.

Some plants, such as sugar beet, potatoes and pears, contain pectins with acetylated galacturonic acid in addition to methyl esters. Acetylation prevents gel-formation but increases the stabilising and emulsifying effects of pectin.

Amidated pectin is a modified form of pectin. Here, some of the galacturonic acid is converted with ammonia to carboxylic acid amide. These pectins are more tolerant of varying calcium concentrations that occur in use.

To prepare a pectin-gel, the ingredients are heated, dissolving the pectin. Upon cooling below gelling temperature, a gel starts to form. If gel formation is too strong, syneresis or a granular texture are the result, while weak gelling leads to excessively soft gels.

Amidated pectins behave like low-ester pectins but need less calcium and are more tolerant of excess calcium. Also, gels from amidated pectin are thermo-reversible; they can be heated and after cooling solidify again, whereas conventional pectin-gels will afterwards remain liquid.

High-ester pectins set at higher temperatures than low-ester pectins. However, gelling reactions with calcium increase as the degree of esterification falls. Similarly, lower pH-values or higher soluble solids (normally sugars) increase gelling speeds. Suitable pectins can therefore be selected for jams and jellies, or for higher-sugar confectionery jellies.

Sources and Production

Pears, apples, guavas, quince, plums, gooseberries, and oranges and other citrus fruits contain large amounts of pectin, while soft fruits, like cherries, grapes, and strawberries, contain small amounts of pectin.

Typical levels of pectin in fresh fruits vegetables are:

- Apples, 1–1.5%

- Apricots, 1%

- Cherries, 0.4%

- Oranges, 0.5–3.5%,

- Carrots 1.4%,

- Citrus peels, 30%.

The main raw materials for pectin production are dried citrus peels or apple pomace, both by-products of juice production. Pomace from sugar beets is also used to a small extent.

From these materials, pectin is extracted by adding hot dilute acid at pH-values from 1.5 – 3.5. During several hours of extraction, the protopectin loses some of its branching and chain length and goes into solution. After filtering, the extract is concentrated in a vacuum and the pectin is then precipitated by adding ethanol or isopropanol. An old technique of precipitating pectin with aluminium salts is no longer used (apart from alcohols and polyvalent cations, pectin also precipitates with proteins and detergents).

Alcohol-precipitated pectin is then separated, washed and dried. Treating the initial pectin with dilute acid leads to low-esterified pectins. When this process includes ammonium hydroxide ($NH_3(aq)$), amidated pectins are obtained. After drying and milling, pectin is usually standard with sugar and sometimes calcium salts or organic acids to have optimum performance in a particular application.

Uses

The main use for pectin is as a gelling agent, thickening agent and stabilizer in food. The classical application is giving the jelly-like consistency to jams or marmalades, which would otherwise be sweet juices. Pectin also reduces syneresis in jams and marmalades and increases the gel strength of low-calorie jams. For household use, pectin is an ingredient in gelling sugar (also known as "jam sugar") where it is diluted to the right concentration with sugar and some citric acid to adjust pH. In some countries, pectin is also available as a solution or an extract, or as a blended powder, for home jam making.

For conventional jams and marmalades that contain above 60% sugar and soluble fruit solids, high-ester pectins are used. With low-ester pectins and amidated pectins, less sugar is needed, so that diet products can be made. Water extract of aiyu seeds is traditionally used in Taiwan to make aiyu jelly, where the extract gels without heating due to low-ester pectins from the seeds and the bivalent cations from the water.

Pectin is used in confectionery jellies to give a good gel structure, a clean bite and to confer a good flavour release. Pectin can also be used to stabilize acidic protein drinks, such as drinking yogurt, to improve the mouth-feel and the pulp stability in juice based drinks and as a fat substitute in baked goods. Typical levels of pectin used as a food additive are between 0.5 and 1.0% – this is about the same amount of pectin as in fresh fruit.

In medicine, pectin increases viscosity and volume of stool so that it is used against constipation and diarrhea. Until 2002, it was one of the main ingredients used in Kaopectate a medication to combat diarrhea, along with kaolinite. It has been used in gentle heavy metal removal from biological systems. Pectin is also used in throat lozenges as a demulcent.

In cosmetic products, pectin acts as a stabilizer. Pectin is also used in wound healing preparations and specialty medical adhesives, such as colostomy devices.

Sriamornsak revealed that pectin could be used in various oral drug delivery platforms, e.g., controlled release systems, gastro-retentive systems, colon-specific delivery systems and mucoadhesive delivery systems, according to its intoxicity and low cost. It was found that pectin from different sources provides different gelling abilities, due to variations in molecular size and chemical composition. Like other natural polymers, a major problem with pectin is inconsistency in reproducibility between samples, which may result in poor reproducibility in drug delivery characteristics.

In ruminant nutrition, depending on the extent of lignification of the cell wall, pectin is up to 90% digestible by bacterial enzymes. Ruminant nutritionists recommend that the digestibility and energy concentration in forages be improved by increasing pectin concentration in the forage.

In cigars, pectin is considered an excellent substitute for vegetable glue and many cigar smokers and collectors use pectin for repairing damaged tobacco leaves on their cigars.

Yablokov et al., writing in Chernobyl: Consequences of the Catastrophe for People and the Environment, quote research conducted by the Ukrainian Center of Radiation Medicine and the Belarusian Institute of Radiation Medicine and Endocrinology, concluded, regarding pectin's radioprotective effects, that "adding pectin preparations to the food of inhabitants of the Chernobyl-contaminated regions promotes an effective excretion of incorporated radionuclides" such as cesium-137. The positive results of using pectin food additive preparations in a number of clinical studies conducted on children in severely polluted areas, with up to 50% improvement over control groups.

During the Second World War, Allied pilots were provided with maps printed on silk, for navigation in escape and evasion efforts. The printing process at first proved nearly impossible because the several layers of ink immediately ran, blurring outlines and rendering place names illegible until the inventor of the maps, Clayton Hutton, mixed a little pectin with the ink and at once the pectin coagulated the ink and prevented it from running, allowing small topographic features to be clearly visible.

Plasmodesmata

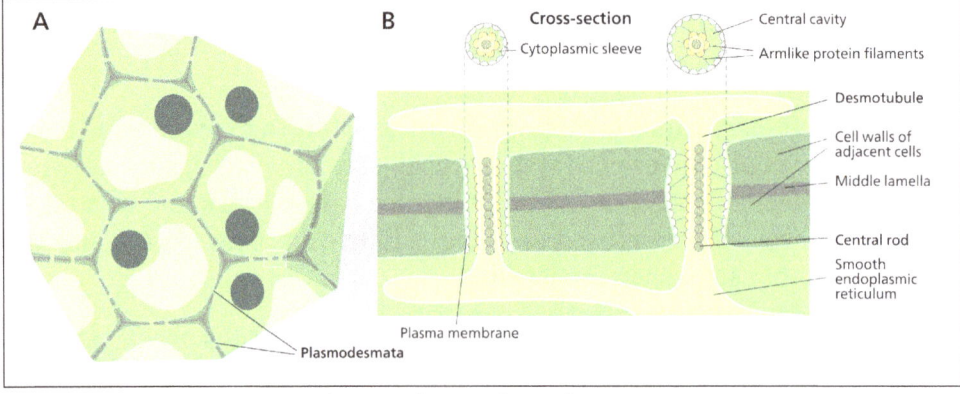

Diagram of some plasmodesmata.

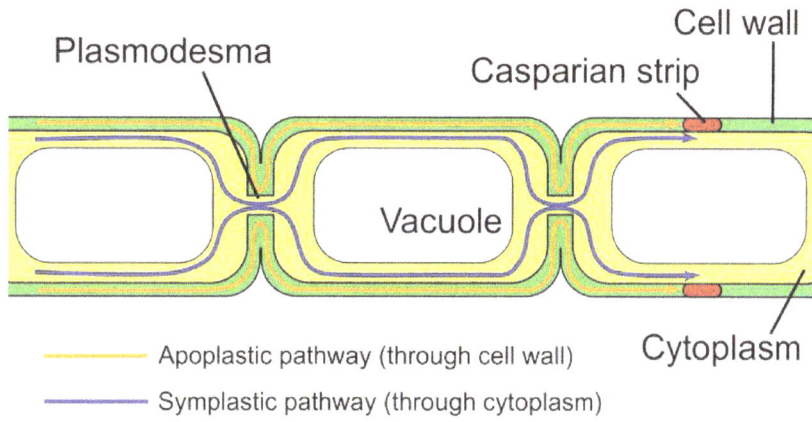

Plasmodesmata allow molecules to travel between plant cells through the symplastic pathway.

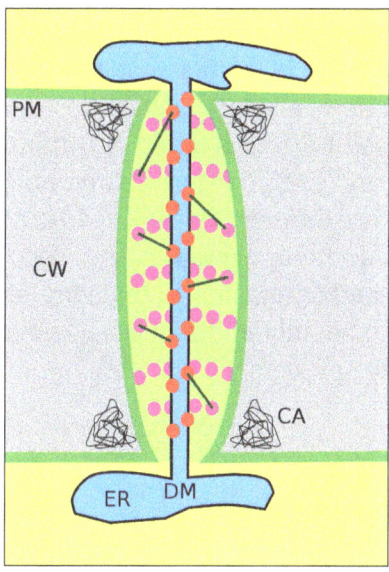

The structure of a primary plasmodesma. CW=Cell wall CA=Callose PM=Plasma membrane ER=Endoplasmic reticulum DM=Desmotubule Red circles=Actin Purple circles and spokes=Other unidentified proteins.

Plasmodesmata (singular: plasmodesma) are microscopic channels which traverse the cell walls of plant cells and some algal cells, enabling transport and communication between them. Plasmodesmata evolved independently in several lineages, and species that have these structures include members of the Charophyceae, Charales, Coleochaetales and Phaeophyceae (which are all algae), as well as all embryophytes, better known as land plants. Unlike animal cells, almost every plant cell is surrounded by a polysaccharide cell wall. Neighbouring plant cells are therefore separated by a pair of cell walls and the intervening middle lamella, forming an extracellular domain known as the apoplast. Although cell walls are permeable to small soluble proteins and other solutes, plasmodesmata enable direct, regulated, symplastic transport of substances between cells. There are two forms of plasmodesmata: primary plasmodesmata, which are formed during cell division, and secondary plasmodesmata, which can form between mature cells.

Similar structures, called gap junctions and membrane nanotubes, interconnect animal cells and stromules form between plastids in plant cells.

Formation

Primary plasmodesmata are formed when fractions of the endoplasmic reticulum are trapped across the middle lamella as new cell wall are synthesized between two newly divided plant cells. These eventually become the cytoplasmic connections between cells. At the formation site, the wall is not thickened further, and depressions or thin areas known as pits are formed in the walls. Pits normally pair up between adjacent cells. Plasmodesmata can also be inserted into existing cell walls between non-dividing cells (secondary plasmodesmata).

Primary Plasmodesmata

The formation of primary plasmodesmata occurs during the part of the cellular division process where the endoplasmic reticulum and the new plate are fused together, this process results in the formation of a cytoplasmic pore (or cytoplasmic sleeve). The desmotubule, also known as the appressed ER, forms alongside the cortical ER. Both the appressed ER and the cortical ER are packed tightly together, thus leaving no room for any luminal space. It is proposed that the appressed ER acts as a membrane transportation route in the plasmodesmata. When filaments of the cortical ER are entangled in the formation of a new cell plate, plasmodesmata formation occurs in land plants. It is hypothesized that the appressed ER forms due to a combination of pressure from a growing cell wall and interaction from ER and PM proteins. Primary plasmodesmata are often present in areas where the cell walls appear to be thinner. This is due to the fact that as a cell wall expands, the abundance of the primary plasmodesmata decreases. In order to further expand plasmodesmal density during cell wall growth secondary plasmodesmata are produced. The process of secondary plasmodesmata formation is still to be fully understood, however various degrading enzymes and ER proteins are said to stimulate the process.

Structure

Plasmodesmatal Plasma Membrane

A typical plant cell may have between 10^3 and 10^5 plasmodesmata connecting it with adjacent cells equating to between 1 and 10 per μm^2. Plasmodesmata are approximately 50–60 nm in diameter at the midpoint and are constructed of three main layers, the plasma membrane, the cytoplasmic sleeve, and the desmotubule. They can transverse cell walls that are up to 90 nm thick.

The plasma membrane portion of the plasmodesma is a continuous extension of the cell membrane or plasmalemma and has a similar phospholipid bilayer structure.

The cytoplasmic sleeve is a fluid-filled space enclosed by the plasmalemma and is a continuous extension of the cytosol. Trafficking of molecules and ions through plasmodesmata occurs through this space. Smaller molecules (e.g. sugars and amino acids) and ions can easily pass through plasmodesmata by diffusion without the need for additional chemical energy. Larger molecules, including proteins (for example green fluorescent protein) and RNA, can also pass through the cytoplasmic sleeve diffusively. Plasmodesmatal transport of some larger molecules is facilitated by mechanisms that are currently unknown. One mechanism of regulation of the permeability of plasmodesmata is the accumulation of the polysaccharide callose around the neck region to form a collar, thereby reducing the diameter of the pore available for transport of substances.

Through dilation, active gating or structural remodeling the permeability of the plasmodesmata is increased. This increase in plasmodesmata pore permeability allows for larger molecules, or ((macromolecules)), such as signaling molecules, transcription factors and RNA-protein complexes to be transported to various cellular compartments.

Desmotubule

The desmotubule is a tube of appressed (flattened) endoplasmic reticulum that runs between two adjacent cells. Some molecules are known to be transported through this channel, but it is not thought to be the main route for plasmodesmatal transport.

Around the desmotubule and the plasma membrane areas of an electron dense material have been seen, often joined together by spoke-like structures that seem to split the plasmodesma into smaller channels. These structures may be composed of myosin and actin, which are part of the cell's cytoskeleton. If this is the case these proteins could be used in the selective transport of large molecules between the two cells.

Transport

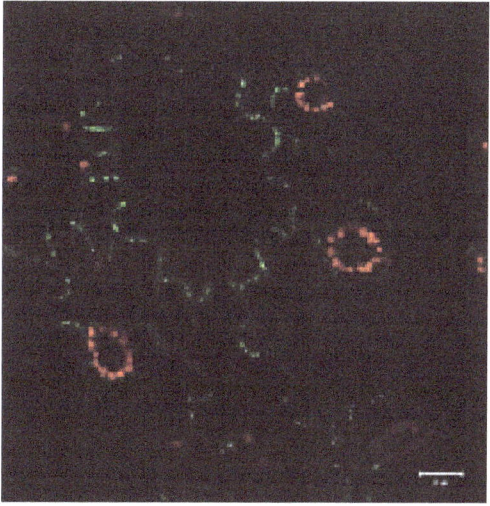

Tobacco mosaic virus movement protein 30 localizes to plasmodesmata.

Plasmodesmata have been shown to transport proteins (including transcription factors), short interfering RNA, messenger RNA, viroids, and viral genomes from cell to cell. One example of a viral movement proteins is the tobacco mosaic virus MP-30. MP-30 is thought to bind to the virus's own genome and shuttle it from infected cells to uninfected cells through plasmodesmata. Flowering Locus T protein moves from leaves to the shoot apical meristem through plasmodesmata to initiate flowering.

Plasmodesmata are also used by cells in phloem, and symplastic transport is used to regulate the sieve-tube cells by the companion cells.

The size of molecules that can pass through plasmodesmata is determined by the size exclusion limit. This limit is highly variable and is subject to active modification. For example, MP-30 is able to increase the size exclusion limit from 700 Daltons to 9400 Daltons thereby aiding its movement through a plant. Also, increasing calcium concentrations in the cytoplasm, either by injection or by cold-induction, has been shown to constrict the opening of surrounding plasmodesmata and limit transport.

Several models for possible active transport through plasmodesmata exist. It has been suggested that such transport is mediated by interactions with proteins localized on the desmotubule, and/or by chaperones partially unfolding proteins, allowing them to fit through the narrow passage. A similar mechanism may be involved in transporting viral nucleic acids through the plasmodesmata.

Cytoskeletal Components of Plasmodesmata

Plasmodesmata link almost every cell within a plant, which can cause negative effects such as the spread of viruses. In order to understand this we must first look at cytoskeletal components, such as actin microfilaments, microtubules, and myosin proteins, and how they are related to cell to cell transport. Actin microfilaments are linked to the transport of viral movement proteins to plasmodesmata which allow for cell to cell transport through the plasmodesmata. Fluorescent tagging for co-expression in tobacco leaves showed that actin filaments are responsible for transporting viral movement proteins to the plasmodesmata. When actin polymerization was blocked it caused a decrease in plasmodesmata targeting of the movement proteins in the tobacco and allowed for 10-kDa (rather than 126-kDa) components to move between tobacco mesophyll cells. This also impacted cell to cell movement of molecules within the tobacco plant.

Viruses

Viruses break down actin filaments within the plasmodesmata channel in order to move within the plant. For example, when the cucumber mosaic virus (CMV) gets into plants it is able to travel through almost every cell through utilization of viral movement proteins to transport themselves through the plasmodesmata. When tobacco leaves are treated with a drug that stabilizes actin filaments, phalloidin, the cucumber mosaic virus movement proteins are unable to increase the plasmodesmata size exclusion limit (SEL).

Myosin

High amounts of myosin proteins are found at the sights of plasmodesmata. These proteins are involved in directing viral cargoes to plasmodesmata. When mutant forms of myosin were tested in tobacco plants, viral protein targeting to plasmodesmata was negatively affected. Permanent binding of myosin to actin, which was induced by a drug, caused a decrease in cell to cell movement. Viruses are also able to selectively bind to myosin proteins.

Microtubules

Microtubules are also are also an important role in cell to cell transport of viral RNA. Viruses use many different methods of transporting themselves from cell to cell, and one of those methods associating the N-terminal domain of its RNA to localize to plasmodesmata through microtubules. Tobacco plants injected with tobacco movement viruses that were kept in high temperatures there was a strong correlation between TMV movement proteins that were attached to GFP with microtubules. This lead to an increase in the spread of viral RNA through the tobacco.

Plasmodesmata and Callose

Plasmodesmata regulation and structure are regulated by a beta 1,3-glucan polymer known as callose. Callose is found in cell plates during the process of cytokinesis, as this process reaches completion the levels of calls decrease. The only callose rich parts of the cell include the sections of the cell wall that plasmodesmata are present. In order to regulate what is transported in the plasmodesmata, callose must be present. Callose provides the mechanism in which plasmodesmata permeability is regulated. In order to control what is transported between different tissues, the plasmodesmata undergo several specialized conformational changes.

The activity of plasmodesmata are linked to physiological and developmental processes within plants. There is a hormone signaling pathway that relays primary cellular signals to the plasmodesmata. There are also patterns of environmental, physiological, and developmental cues that show relation to plasmodesmata function. An important mechanism of plasmodesmata is the ability to gate its channels. Callose levels have been proved to be a method of changing plasmodesmata aperture size. Callose deposits are found at the neck of the plasmodesmata in new cell walls that have been formed. The level of deposits at the plasmodesmata can fluctuate which shows that there are signals that trigger an accumulation of callose at the plasmodesmata and cause plasmodesmata to become gated or more open. Enzyme activities of Beta 1,3-glucan synthase and hydrolases are involved in changes in plasmodesmata cellulose level. Some extracellular signals change transcription of activities of this synthase and hydrolase. Arabidopsis thailana contain callose synthase genes that encode a catalytic subunit of B-1,3-glucan. Gain of function mutants in this gene pool show increased deposition of callose at plasmodesmata and a decrease in macromolecular trafficking as well as a defective root system during development.

Middle Lamella

The middle lamella is a pectin layer which functions to cement the two adjoining cells together of the cell wall. This is essential to plants as it gives them stability, and allows that plants can form plasmodesmata between cells. The middle lamella is the first layer that is formed, which is deposited at the time of cytokinesis. The cell plate that is molded during cell division is developed into lamellum or the middle lamella. This layer is basically made up of calcium and magnesium pectates.

The pectins in plants form a continuous and unified layer between adjacent cells. Most of the time, the middle lamella is quite difficult to distinguish from the primary cell wall, especially in certain cells that develop a thick secondary wall. In cases like this, when the middle lamella and the two adjacent primary walls, and even the first layer of the secondary wall of every cell, is normally referred as a compound middle lamella. For the multicellular organisms, when the middle lamella dissolves, the cells will be isolated from each other. This happens when leaves and petals fall.

Cell Membrane

The cell membrane is also known as the plasma membrane (PM) or cytoplasmic membrane, and historically referred to as the plasmalemma. It is a biological membrane that separates the interior of all cells from the outside environment (the extracellular space) which protects the cell from its environment. Cell membrane consists of a lipid bilayer, including cholesterols (a lipid component) that sit between phospholipids to maintain their fluidity under various temperature, in combination with proteins such as integral proteins, and peripheral proteins that go across inside and outside of the membrane serving as membrane transporter, and loosely attached to the outer (peripheral) side of the cell membrane acting as several kinds of enzymes shaping the cell, respectively.

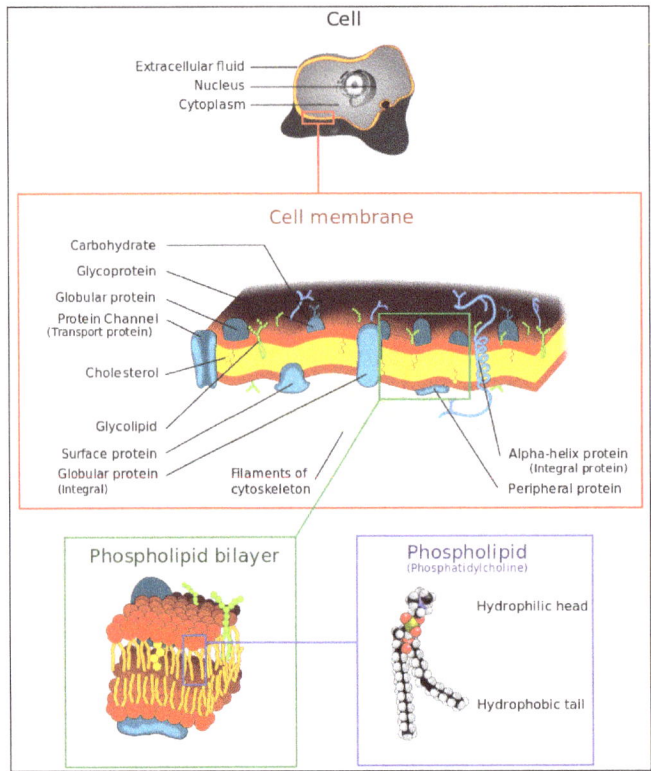

Illustration of a Eukaryotic cell membrane.

The cell membrane controls the movement of substances in and out of cells and organelles. In this way, it is selectively permeable to ions and organic molecules. In addition, cell membranes are involved in a variety of cellular processes such as cell adhesion, ion conductivity and cell signalling and serve as the attachment surface for several extracellular structures, including the cell wall, the carbohydrate layer called the glycocalyx, and the intracellular network of protein fibers called the cytoskeleton. In the field of synthetic biology, cell membranes can be artificially reassembled.

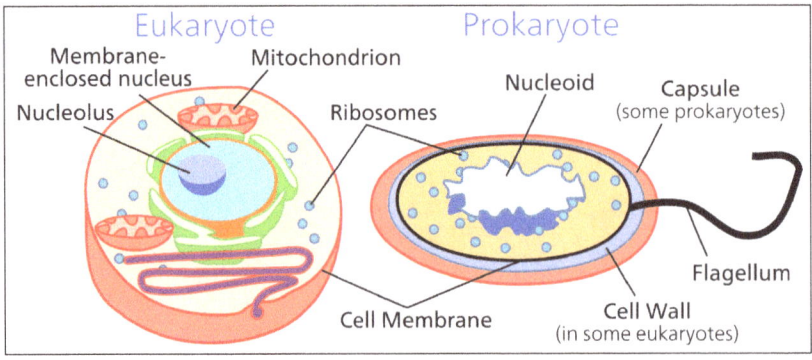

Comparison of Eukaryotes vs. Prokaryotes.

Composition

Cell membranes contain a variety of biological molecules, notably lipids and proteins. Composition is not set, but constantly changing for fluidity and changes in the environment, even fluctuating during different stages of cell development. Specifically, the amount of cholesterol in human primary neuron cell membrane changes, and this change in composition affects fluidity throughout development stages.

Material is incorporated into the membrane, or deleted from it, by a variety of mechanisms:

- Fusion of intracellular vesicles with the membrane (exocytosis) not only excretes the contents of the vesicle but also incorporates the vesicle membrane's components into the cell membrane. The membrane may form blebs around extracellular material that pinch off to become vesicles (endocytosis).

- If a membrane is continuous with a tubular structure made of membrane material, then material from the tube can be drawn into the membrane continuously.

- Although the concentration of membrane components in the aqueous phase is low (stable membrane components have low solubility in water), there is an exchange of molecules between the lipid and aqueous phases.

Lipids

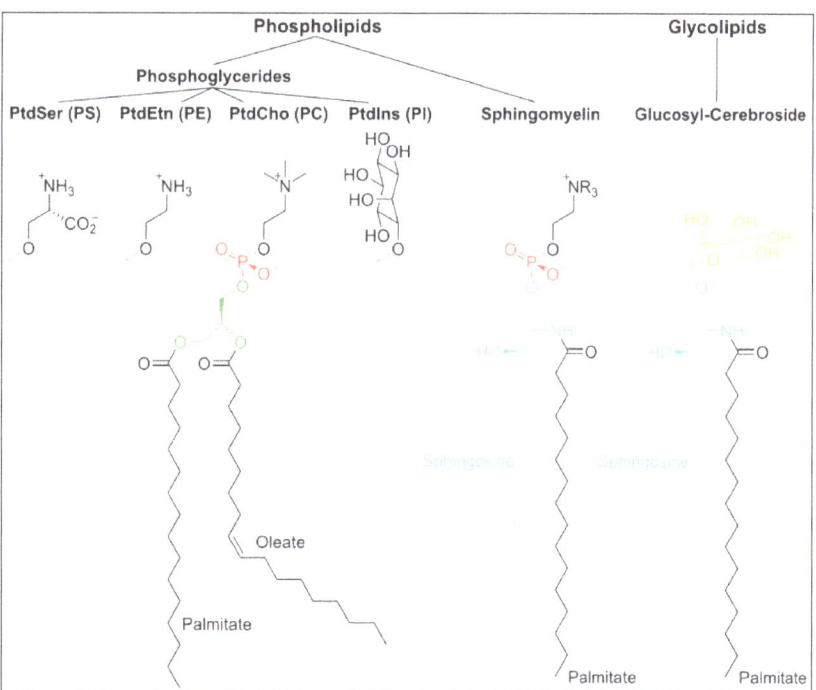

Examples of the major membrane phospholipids and glycolipids: phosphatidylcholine (PtdCho), phosphatidylethanolamine (PtdEtn), phosphatidylinositol (PtdIns), phosphatidylserine (PtdSer).

The cell membrane consists of three classes of amphipathic lipids: phospholipids, glycolipids, and sterols. The amount of each depends upon the type of cell, but in the majority of cases phospholipids are the most abundant, often contributing for over 50% of all lipids in plasma membranes.

Glycolipids only account for a minute amount of about 2% and sterols make up the rest. In RBC studies, 30% of the plasma membrane is lipid. However, for the majority of eukaryotic cells, the composition of plasma membranes is about half lipids and half proteins by weight.

The fatty chains in phospholipids and glycolipids usually contain an even number of carbon atoms, typically between 16 and 20. The 16- and 18-carbon fatty acids are the most common. Fatty acids may be saturated or unsaturated, with the configuration of the double bonds nearly always "cis". The length and the degree of unsaturation of fatty acid chains have a profound effect on membrane fluidity as unsaturated lipids create a kink, preventing the fatty acids from packing together as tightly, thus decreasing the melting temperature (increasing the fluidity) of the membrane. The ability of some organisms to regulate the fluidity of their cell membranes by altering lipid composition is called homeoviscous adaptation.

The entire membrane is held together via non-covalent interaction of hydrophobic tails, however the structure is quite fluid and not fixed rigidly in place. Under physiological conditions phospholipid molecules in the cell membrane are in the liquid crystalline state. It means the lipid molecules are free to diffuse and exhibit rapid lateral diffusion along the layer in which they are present. However, the exchange of phospholipid molecules between intracellular and extracellular leaflets of the bilayer is a very slow process. Lipid rafts and caveolae are examples of cholesterol-enriched microdomains in the cell membrane. Also, a fraction of the lipid in direct contact with integral membrane proteins, which is tightly bound to the protein surface is called annular lipid shell; it behaves as a part of protein complex.

In animal cells cholesterol is normally found dispersed in varying degrees throughout cell membranes, in the irregular spaces between the hydrophobic tails of the membrane lipids, where it confers a stiffening and strengthening effect on the membrane. Additionally, the amount of cholesterol in biological membranes varies between organisms, cell types, and even in individual cells. Cholesterol, a major component of animal plasma membranes, regulates the fluidity of the overall membrane, meaning that cholesterol controls the amount of movement of the various cell membrane components based on its concentrations. In high temperatures, cholesterol inhibits the movement of phospholipid fatty acid chains, causing a reduced permeability to small molecules and reduced membrane fluidity. The opposite is true for the role of cholesterol in cooler temperatures. Cholesterol production, and thus concentration, is up-regulated (increased) in response to cold temperature. At cold temperatures, cholesterol interferes with fatty acid chain interactions. Acting as antifreeze, cholesterol maintains the fluidity of the membrane. Cholesterol is more abundant in cold-weather animals than warm-weather animals. In plants, which lack cholesterol, related compounds called sterols perform the same function as cholesterol.

Phospholipids Forming Lipid Vesicles

Lipid vesicles or liposomes are approximately spherical pockets that are enclosed by a lipid bilayer. These structures are used in laboratories to study the effects of chemicals in cells by delivering these chemicals directly to the cell, as well as getting more insight into cell membrane permeability. Lipid vesicles and liposomes are formed by first suspending a lipid in an aqueous solution then agitating the mixture through sonication, resulting in a vesicle. By measuring the rate of efflux from that of the inside of the vesicle to the ambient solution, allows researcher to better understand membrane permeability. Vesicles can be formed with molecules and ions inside the vesicle by forming the vesicle with the desired molecule or ion present in the solution. Proteins can also be embedded into the membrane through solubilizing the desired proteins in the presence of detergents and attaching them to the phospholipids in which the liposome is formed. These provide researchers with a tool to examine various membrane protein functions.

Carbohydrates

Plasma membranes also contain carbohydrates, predominantly glycoproteins, but with some glycolipids (cerebrosides and gangliosides). Carbohydrates are important in the role of cell-cell recognition in eukaryotes; they are located on the surface of the cell where they recognize host cells and share information, viruses that bind to cells using these receptors cause an infection For the most part, no glycosylation occurs on membranes within the cell; rather generally glycosylation occurs on the extracellular surface of the plasma membrane. The glycocalyx is an important feature in all cells, especially epithelia with microvilli. Recent data suggest the glycocalyx participates in cell adhesion, lymphocyte homing, and many others. The penultimate sugar is galactose and the terminal sugar is sialic acid, as the sugar backbone is modified in the Golgi apparatus. Sialic acid carries a negative charge, providing an external barrier to charged particles.

Proteins

Type	Description	Examples
Integral proteins or transmembrane proteins	Span the membrane and have a hydrophilic cytosolic domain, which interacts with internal molecules, a hydrophobic membrane-spanning domain that anchors it within the cell membrane, and a hydrophilic extracellular domain that interacts with external molecules. The hydrophobic domain consists of one, multiple, or a combination of α-helices and β sheet protein motifs.	Ion channels, proton pumps, G protein-coupled receptor
Lipid anchored proteins	Covalently bound to single or multiple lipid molecules; hydrophobically insert into the cell membrane and anchor the protein. The protein itself is not in contact with the membrane.	G proteins
Peripheral proteins	Attached to integral membrane proteins, or associated with peripheral regions of the lipid bilayer. These proteins tend to have only temporary interactions with biological membranes, and once reacted, the molecule dissociates to carry on its work in the cytoplasm.	Some enzymes, some hormones

The cell membrane has large content of proteins, typically around 50% of membrane volume These proteins are important for cell because they are responsible for various biological activities. Approximately a third of the genes in yeast code specifically for them, and this number is even higher in multicellular organisms. Membrane proteins consist of three main types: Integral proteins, peripheral proteins, and lipid-anchored proteins.

As shown in the adjacent table, integral proteins are amphipathic transmembrane proteins. Examples of integral proteins include ion channels, proton pumps, and g-protein coupled receptors. Ion channels allow inorganic ions such as sodium, potassium, calcium, or chlorine to diffuse down their electrochemical gradient across the lipid bilayer through hydrophilic pores across the membrane. The electrical behavior of cells (i.e. nerve cells) are controlled by ion channels. Proton pumps are protein pumps that are embedded in the lipid bilayer that allow protons to travel through the membrane by transferring from one amino acid side chain to another. Processes such as electron transport and generating ATP use proton pumps. A G-protein coupled receptor is a single polypeptide chain that crosses the lipid bilayer seven times responding to signal molecules (i.e. hormones and neurotransmitters). G-protein coupled receptors are used in processes such as cell to cell signaling, the regulation of the production of cAMP, and the regulation of ion channels.

The cell membrane, being exposed to the outside environment, is an important site of cell–cell communication. As such, a large variety of protein receptors and identification proteins, such as antigens, are present on the surface of the membrane. Functions of membrane proteins can also include cell–cell contact, surface recognition, cytoskeleton contact, signaling, enzymatic activity, or transporting substances across the membrane.

Most membrane proteins must be inserted in some way into the membrane. For this to occur, an N-terminus "signal sequence" of amino acids directs proteins to the endoplasmic reticulum, which inserts the proteins into a lipid bilayer. Once inserted, the proteins are then transported to their final destination in vesicles, where the vesicle fuses with the target membrane.

Function

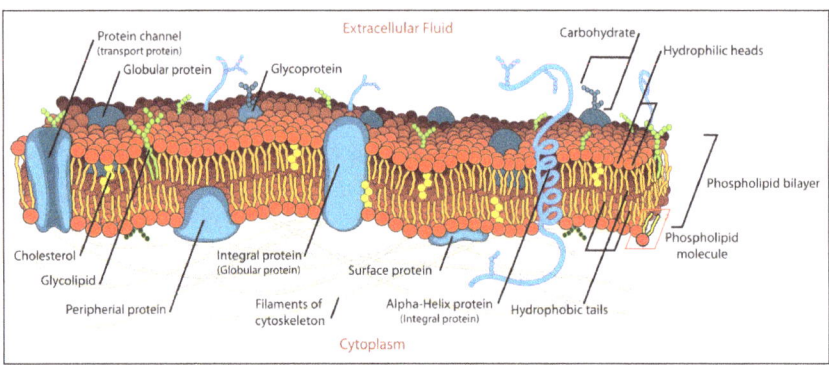

A detailed diagram of the cell membrane.

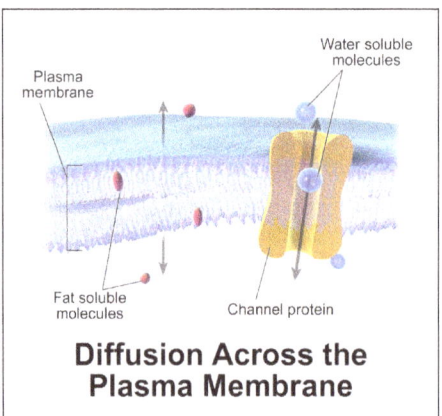

Illustration depicting cellular diffusion.

The cell membrane surrounds the cytoplasm of living cells, physically separating the intracellular components from the extracellular environment. The cell membrane also plays a role in anchoring the cytoskeleton to provide shape to the cell, and in attaching to the extracellular matrix and other cells to hold them together to form tissues. Fungi, bacteria, most archaea, and plants also have a cell wall, which provides a mechanical support to the cell and precludes the passage of larger molecules.

The cell membrane is selectively permeable and able to regulate what enters and exits the cell, thus facilitating the transport of materials needed for survival. The movement of substances across the membrane can be either "passive", occurring without the input of cellular energy, or "active", requiring the cell to expend energy in transporting it. The membrane also maintains the cell potential. The cell membrane thus works as a selective filter that allows only certain things to come inside or go outside the cell. The cell employs a number of transport mechanisms that involve biological membranes:

1. Passive osmosis and diffusion: Some substances (small molecules, ions) such as carbon dioxide (CO_2) and oxygen (O_2), can move across the plasma membrane by diffusion, which is a passive transport process. Because the membrane acts as a barrier for certain molecules and ions, they can occur in different concentrations on the two sides of the membrane. Diffusion occurs when small molecules and ions move freely from high concentration to low concentration in order to equilibrate the membrane. It is considered a passive transport process because it does not require energy and is propelled by the concentration gradient created by each side of the membrane. Such a concentration gradient across a semipermeable membrane sets up an osmotic flow for the water. Osmosis, in biological systems involves a solvent, moving through a semipermeable membrane similarly to passive diffusion as the solvent still moves with the concentration gradient and requires no energy. While water is the most common solvent in cell, it can also be other liquids as well as supercritical liquids and gases.

2. Transmembrane protein channels and transporters: Transmembrane proteins extend through the lipid bilayer of the membranes; they function on both sides of the membrane to transport molecules across it. Nutrients, such as sugars or amino acids, must enter the cell, and certain products of metabolism must leave the cell. Such molecules can diffuse passively through protein channels such as aquaporins in facilitated diffusion or are pumped across the membrane by transmembrane transporters. Protein channel proteins, also called permeases, are usually quite specific, and they only recognize and transport a limited variety of chemical substances, often limited to a single substance. Another example of a transmembrane protein is a cell-surface receptor, which allow cell signaling molecules to communicate between cells.

3. Endocytosis: Endocytosis is the process in which cells absorb molecules by engulfing them. The plasma membrane creates a small deformation inward, called an invagination, in which the substance to be transported is captured. This invagination is caused by proteins on the outside on the cell membrane, acting as receptors and clustering into depressions that eventually promote accumulation of more proteins and lipids on the cytosolic side of the membrane. The deformation then pinches off from the membrane on the inside of the cell, creating a vesicle containing the captured substance. Endocytosis is a pathway for internalizing solid particles ("cell eating" or phagocytosis), small molecules and ions ("cell drinking" or pinocytosis), and macromolecules. Endocytosis requires energy and is thus a form of active transport.

4. Exocytosis: Just as material can be brought into the cell by invagination and formation of a vesicle, the membrane of a vesicle can be fused with the plasma membrane, extruding its contents to the surrounding medium. This is the process of exocytosis. Exocytosis occurs in various cells to

remove undigested residues of substances brought in by endocytosis, to secrete substances such as hormones and enzymes, and to transport a substance completely across a cellular barrier. In the process of exocytosis, the undigested waste-containing food vacuole or the secretory vesicle budded from Golgi apparatus, is first moved by cytoskeleton from the interior of the cell to the surface. The vesicle membrane comes in contact with the plasma membrane. The lipid molecules of the two bilayers rearrange themselves and the two membranes are, thus, fused. A passage is formed in the fused membrane and the vesicles discharges its contents outside the cell.

Prokaryotes

Prokaryotes are divided into two different groups, Archaea and Bacteria, with bacteria dividing further into gram-positive and gram-negative. Gram-negative bacteria have both a plasma membrane and an outer membrane separated by periplasm, however, other prokaryotes have only a plasma membrane. These two membranes differ in many aspects. The outer membrane of the gram-negative bacteria differ from other prokaryotes due to phospholipids forming the exterior of the bilayer, and lipoproteins and phospholipids forming the interior. The outer membrane typically has a porous quality due to its presence of membrane proteins, such as gram-negative porins, which are pore-forming proteins. The inner, plasma membrane is also generally symmetric whereas the outer membrane is asymmetric because of proteins such as the aforementioned. Also, for the prokaryotic membranes, there are multiple things that can affect the fluidity. One of the major factors that can affect the fluidity is fatty acid composition. For example, when the bacteria Staphylococcus aureus was grown in 37°C for 24h, the membrane exhibited a more fluid state instead of a gel-like state. This supports the concept that in higher temperatures, the membrane is more fluid than in colder temperatures. When the membrane is becoming more fluid and needs to become more stabilized, it will make longer fatty acid chains or saturated fatty acid chains in order to help stabilize the membrane. Bacteria are also surrounded by a cell wall composed of peptidoglycan (amino acids and sugars). Some eukaryotic cells also have cell walls, but none that are made of peptidoglycan. The outer membrane of gram negative bacteria is rich in lipopolysaccharides, which are combined poly- or oligosaccharide and carbohydrate lipid regions that stimulate the cell's natural immunity. The outer membrane can bleb out into periplasmic protrusions under stress conditions or upon virulence requirements while encountering a host target cell, and thus such blebs may work as virulence organelles. Bacterial cells provide numerous examples of the diverse ways in which prokaryotic cell membranes are adapted with structures that suit the organism's niche. For example, proteins on the surface of certain bacterial cells aid in their gliding motion. Many gram-negative bacteria have cell membranes which contain ATP-driven protein exporting systems.

Structures

Fluid Mosaic Model

According to the fluid mosaic model of S. J. Singer and G. L. Nicolson (1972), which replaced the earlier model of Davson and Danielli, biological membranes can be considered as a two-dimensional liquid in which lipid and protein molecules diffuse more or less easily. Although the lipid bilayers that form the basis of the membranes do indeed form two-dimensional liquids by themselves, the plasma membrane also contains a large quantity of proteins, which provide more structure. Examples of such structures are protein-protein complexes, pickets and fences formed by the actin-based cytoskeleton, and potentially lipid rafts.

Lipid Bilayer

Diagram of the arrangement of amphipathic lipid molecules to form a lipid bilayer. The yellow polar head groups separate the grey hydrophobic tails from the aqueous cytosolic and extracellular environments.

Lipid bilayers form through the process of self-assembly. The cell membrane consists primarily of a thin layer of amphipathic phospholipids that spontaneously arrange so that the hydrophobic "tail" regions are isolated from the surrounding water while the hydrophilic "head" regions interact with the intracellular (cytosolic) and extracellular faces of the resulting bilayer. This forms a continuous, spherical lipid bilayer. Hydrophobic interactions (also known as the hydrophobic effect) are the major driving forces in the formation of lipid bilayers. An increase in interactions between hydrophobic molecules (causing clustering of hydrophobic regions) allows water molecules to bond more freely with each other, increasing the entropy of the system. This complex interaction can include noncovalent interactions such as van der Waals, electrostatic and hydrogen bonds.

Lipid bilayers are generally impermeable to ions and polar molecules. The arrangement of hydrophilic heads and hydrophobic tails of the lipid bilayer prevent polar solutes (ex. amino acids, nucleic acids, carbohydrates, proteins, and ions) from diffusing across the membrane, but generally allows for the passive diffusion of hydrophobic molecules. This affords the cell the ability to control the movement of these substances via transmembrane protein complexes such as pores, channels and gates. Flippases and scramblases concentrate phosphatidyl serine, which carries a negative charge, on the inner membrane. Along with NANA, this creates an extra barrier to charged moieties moving through the membrane.

Membranes serve diverse functions in eukaryotic and prokaryotic cells. One important role is to regulate the movement of materials into and out of cells. The phospholipid bilayer structure (fluid mosaic model) with specific membrane proteins accounts for the selective permeability of the membrane and passive and active transport mechanisms. In addition, membranes in prokaryotes and in the mitochondria and chloroplasts of eukaryotes facilitate the synthesis of ATP through chemiosmosis.

Membrane Polarity

The apical membrane of a polarized cell is the surface of the plasma membrane that faces inward to the lumen. This is particularly evident in epithelial and endothelial cells, but also describes other polarized cells, such as neurons. The basolateral membrane of a polarized cell is the surface of the plasma membrane that forms its basal and lateral surfaces. It faces outwards, towards the interstitium, and away from the lumen. Basolateral membrane is a compound phrase referring to the terms "basal (base) membrane" and "lateral (side) membrane", which, especially in epithelial cells, are identical in composition and activity. Proteins (such as ion channels and pumps) are free to move from the basal to the lateral surface of the cell or vice versa in accordance with the fluid mosaic model. Tight junctions join epithelial cells near their apical surface to prevent the migration of proteins from the basolateral membrane to the apical membrane. The basal and lateral surfaces thus remain roughly equivalent to one another, yet distinct from the apical surface.

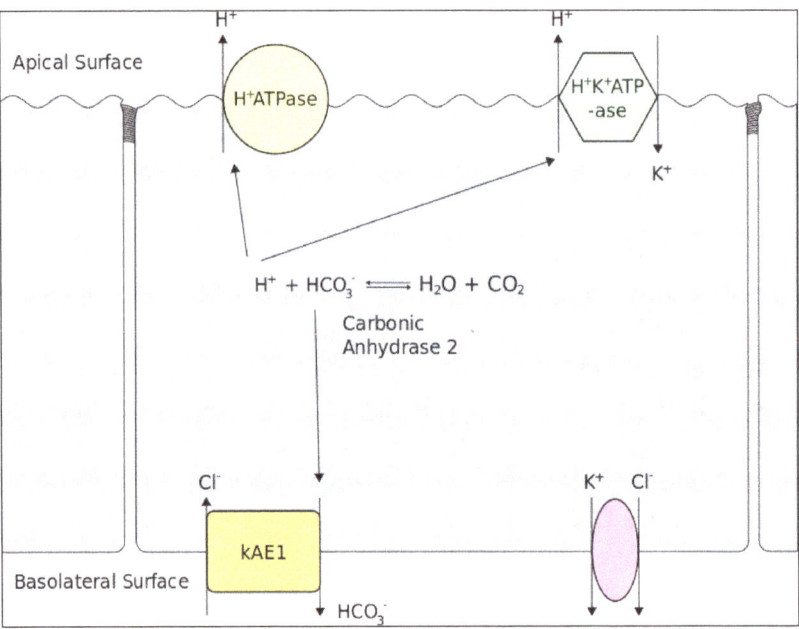

Alpha intercalated cell.

Membrane Structures

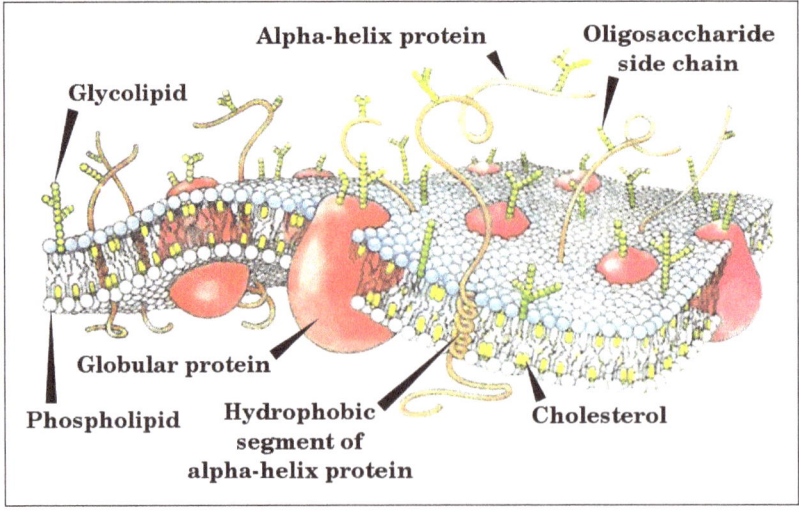

Diagram of the Cell Membrane's structures and their function.

Cell membrane can form different types of "supramembrane" structures such as caveola, post-synaptic density, podosome, invadopodium, focal adhesion, and different types of cell junctions. These structures are usually responsible for cell adhesion, communication, endocytosis and exocytosis. They can be visualized by electron microscopy or fluorescence microscopy. They are composed of specific proteins, such as integrins and cadherins.

Cytoskeleton

The cytoskeleton is found underlying the cell membrane in the cytoplasm and provides a scaffolding for membrane proteins to anchor to, as well as forming organelles that extend from the cell. Indeed, cytoskeletal elements interact extensively and intimately with the cell membrane. Anchoring proteins restricts them to a particular cell surface — for example, the apical surface of epithelial cells that line the vertebrate gut — and limits how far they may diffuse within the bilayer. The cytoskeleton is able to form appendage-like organelles, such as cilia, which are microtubule-based extensions covered by the cell membrane, and filopodia, which are actin-based extensions. These extensions are ensheathed in membrane and project from the surface of the cell in order to sense the external environment and/or make contact with the substrate or other cells. The apical surfaces of epithelial cells are dense with actin-based finger-like projections known as microvilli, which increase cell surface area and thereby increase the absorption rate of nutrients. Localized decoupling of the cytoskeleton and cell membrane results in formation of a bleb.

Intracellular Membranes

The content of the cell, inside the cell membrane, is composed of numerous membrane-bound organelles, which contribute to the overall function of the cell. The origin, structure, and function of each organelle leads to a large variation in the cell composition due to the individual uniqueness associated with each organelle.

- Mitochondria and chloroplasts are considered to have evolved from bacteria, known as the endosymbiotic theory. This theory arose from the idea that Paracoccus and Rhodopseaudomonas, types of bacteria, share similar functions to mitochondria and blue-green algae, or cyanobacteria, share similar functions to chloroplasts. The endosymbiotic theory proposes that through the course of evolution, a eukaryotic cell engulfed these 2 types of bacteria, leading to the formation of mitochondria and chloroplasts inside eukaryotic cells. This engulfment lead to the 2 membranes systems of these organelles in which the outer membrane originated from the host's plasma membrane and the inner membrane was the endosymbiont's plasma membrane. Considering that mitochondria and chloroplasts both contain their own DNA is further support that both of these organelles evolved from engulfed bacteria that thrived inside a eukaryotic cell.

- In eukaryotic cells, the nuclear membrane separates the contents of the nucleus from the cytoplasm of the cell. The nuclear membrane is formed by an inner and outer membrane, providing the strict regulation of materials in to and out of the nucleus. Materials move between the cytosol and the nucleus through nuclear pores in the nuclear membrane. If a cell's nucleus is more active in transcription, its membrane will have more pores. The protein composition of the nucleus can vary greatly from the cytosol as many proteins are unable to cross through pores via diffusion. Within the nuclear membrane, the inner and outer membranes vary in protein composition, and only the outer membrane is continuous with the endoplasmic reticulum (ER) membrane. Like the ER, the outer membrane also possesses ribosomes responsible for producing and transporting proteins into the space between the two membranes. The nuclear membrane disassembles during the early stages of mitosis and reassembles in later stages of mitosis.

• The ER, which is part of the endomembrane system, which makes up a very large portion of the cell's total membrane content. The ER is an enclosed network of tubules and sacs, and its main functions include protein synthesis, and lipid metabolism. There are 2 types of ER, smooth and rough. The rough ER has ribosomes attached to it used for protein synthesis, while the smooth ER is used more for the processing of toxins and calcium regulation in the cell.

• The Golgi apparatus has two interconnected round Golgi cisternae. Compartments of the apparatus forms multiple tubular-reticular networks responsible for organization, stack connection and cargo transport that display a continuous grape-like stringed vesicles ranging from 50-60 nm. The apparatus consists of three main compartments, a flat disc-shaped cisterna with tubular-reticular networks and vesicles.

Variations

The cell membrane has different lipid and protein compositions in distinct types of cells and may have therefore specific names for certain cell types.

• Sarcolemma in myocytes: "Sarcolemma" is the name given to the cell membrane of myocytes (also known as muscle cells). Although the sarcolemma is similar to other cell membranes, it has other functions that set it apart. For instance, the sarcolemma transmits synaptic signals, helps generate action potentials, and is very involved in muscle contractions. Unlike other cell membranes, the sarcolemma makes up small channels called "t-tubules" that pass through the entirety of muscle cells. It has also been found that the average sarcolemma is 10 nm thick as opposed to the 4 nm thickness of a general cell membrane.

• Oolemma is the cell membrane in oocytes: The oolemma of oocytes, (immature egg cells) are not consistent with a lipid bilayer as they lack a bilayer and do not consist of lipids. Rather, the structure has an inner layer, the fertilization envelope, and the exterior is made up of the vitelline layer, which is made up of glycoproteins; however, channels and proteins are still present for their functions in the membrane.

• Axolemma: The specialized plasma membrane on the axons of nerve cells that is responsible for the generation of the action potential. It consists of a granular, densely packed lipid bilayer that works closely with the cytoskeleton components spectrin and actin. These cytoskeleton components are able to bind to and interact with transmembrane proteins in the axolemma.

Permeability

The permeability of a membrane is the rate of passive diffusion of molecules through the membrane. These molecules are known as permeant molecules. Permeability depends mainly on the electric charge and polarity of the molecule and to a lesser extent the molar mass of the molecule. Due to the cell membrane's hydrophobic nature, small electrically neutral molecules pass through the membrane more easily than charged, large ones. The inability of charged molecules to pass through the cell membrane results in pH partition of substances throughout the fluid compartments of the body.

Apoplast

Inside a plant, the apoplast is the space outside the plasma membrane within which material can diffuse freely. It is interrupted by the Casparian strip in roots, by air spaces between plant cells and by the plant cuticle.

Structurally, the apoplast is formed by the continuum of cell walls of adjacent cells as well as the extracellular spaces, forming a tissue level compartment comparable to the symplast. The apoplastic route facilitates the transport of water and solutes across a tissue or organ. This process is known as apoplastic transport.

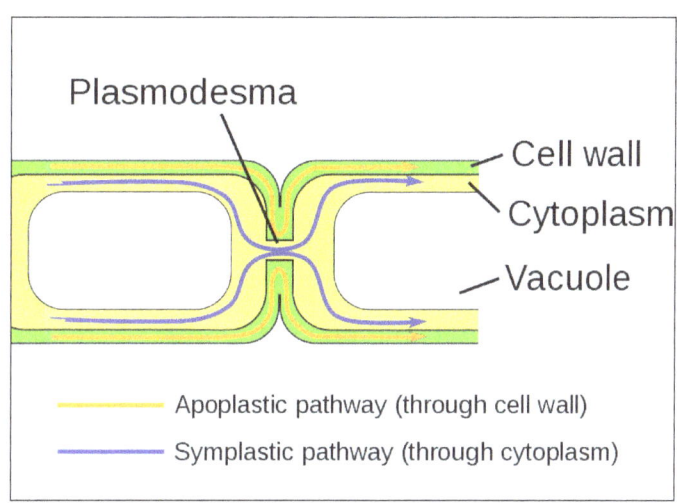

The apoplastic and symplastic pathways.

The apoplast is important for all the plant's interaction with its environment. The main carbon source (carbon dioxide) needs to be solubilized in the apoplast before it diffuses through the plasma membrane into the cell's cytoplasm (symplast) and is used by the chloroplasts during photosynthesis. In the roots, ions diffuse into the apoplast of the epidermis before diffusing into the symplast, or in some cases being taken up by specific ion channels, and being pulled by the plant's transpiration stream, which also occurs completely within the boundaries of the apoplast. Similarly, all gaseous molecules emitted and received by plants such as plant hormones and other pheromones must pass the apoplast. In nitrate poor soils, acidification of the apoplast increases cell wall extensibility and root growth rate. This is believed to be caused by a decrease in nitrate uptake (due to deficit in the soil medium) and supplanted with an increase in chloride uptake. H+ATPase increases the efflux of H+, thus acidifying the apoplast. The apoplast is also a site for cell-to-cell communication. During local oxidative stress, hydrogen peroxide and superoxide anions can diffuse through the apoplast and transport a warning signal to neighbouring cells. In addition, a local alkalinization of the apoplast due to such a stress can travel within minutes to the rest of the plant body via the xylem and trigger systemic acquired resistance. The apoplast also plays an important role in resistance to aluminium toxicity and resistance. Exclusion of aluminium ions in the apoplast prevent toxic levels which inhibit shoot growth, reducing crop yields.

Apoplastic Transport

The apoplastic pathway is one of the two main pathways for water transport in plants, the other being symplastic pathway. In apoplastic transport, water and minerals flow in an upward direction via the apoplast to the xylem in the root. The concentration of solutes transported in aboveground organs is established through a combination of import from the xylem, absorption by cells, and export by the phloem. Transport velocity is higher in the apoplast than the symplast. This method of transport also accounts for a higher proportion of water transport in plant tissues than does symplastic transport. The apoplastic pathway is also involved in passive exclusion. Some of the ions that enter through the roots do not make it to the xylem. The ions are excluded by the plasma membranes of the endodermal cells.

Protoplasm

Protoplasm is defined as the organic and inorganic substances that constitute the living the nucleus, cytoplasm, plastids and mitochondria of the cell.

Protoplasm is a jelly-like substance known to be the living part of the cell. The term was proposed in 1835 and was known as the primary substance responsible for all the living processes.

It was believed that cell were containers of protoplasm. However, the concept could not explain the origin of structures formed within the cell, primarily the nucleus.

Components and Functions of a Protoplasm

The cytoplasm is the first component of protoplasm. It is present between the cell membrane and the nucleus in a eukaryotic cell. All the organelles are found here. It maintains the cell environment. A cytoplasm maintains the shape of cells and also stores substances required by the organelle.

The nucleus is the second component of the protoplasm. The genetic material of an organism is present in the nucleus. Ribosomes are also found in the nucleus, which are essential for the production of proteins in the cell. Prokaryotes contain a nucleoid instead of a nucleus where all the genetic information is found.

Proteins, fats, enzymes, hormones, all make up the protoplasm. These are either dissolved or suspended in the water component of the protoplasm.

Cytoplasm

Cytoplasm has all of the contents in a cell that exist outside of the nucleus that are all encased in the cell membrane inside of the cell. Cytoplasm supports and suspends organelles and cellular molecules while performing processes such as cellular respiration for breathing, synthesizing proteins and having division of cells by both mitosis and meiosis.

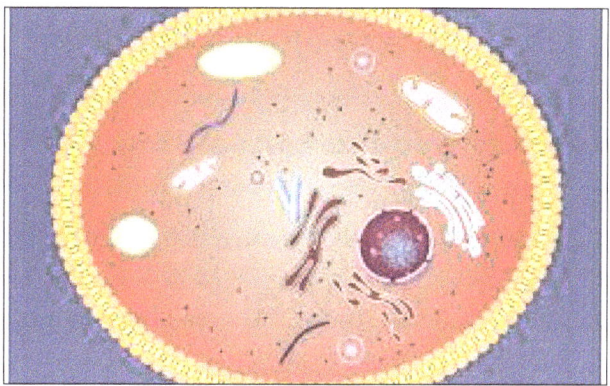

Functions of Cytoplasm

Cytoplasm is a clear substance that is gel-like in the cell membrane but is on the outside of the nucleus. It contains mostly water with the addition of enzymes, organelles, salts and organic molecules. Cytoplasm will liquefy when it is stirred or agitated. It is often referred to as cytosol, meaning "substance of the cell."

Cytoplasm supports and suspends cellular molecules and organelles. Organelles are tiny cellular structures within the cytoplasm that perform specific functions in bacteria or prokaryotic cells and eukaryotic cells of plants, animals and humans. Cytoplasm also helps to move things around in the cells such as hormones and dissolves any cellular waste that may occur.

Cytoplasm moves items around in the cell in a process called cytoplasmic streaming. It also has numerous salts, so it conducts electricity very well. Cytoplasm is also a means of transportation for genetic material in cell division. It is a buffer to protect the genetic material of the cell and keep the organelles from damage when they move and collide with each other. If a cell would be without cytoplasm it could not retain its shape and would be deflated and flat. The organelles would not stay suspended in the solution of a cell without the support of cytoplasm.

Parts of Cytoplasm

Cytoplasm has two main components: the endoplasm and the ectoplasm. The endoplasm is located in the central area of the cytoplasm, and it contains organelles. The ectoplasm is the gel-like substance on the outer portion of the cytoplasm of a cell.

Characteristics of Cytoplasm

Cytoplasm is a heterogeneous mixture of both opaque granules and organic compounds. This combination of these two components gives it the colloidal nature to suspend the organelles in the liquid of the cytoplasm in a cell.

Cytoplasm contains many different shapes and sizes of particles in it and holds them in place in the cell. Cytoplasm contains proteins that are 20 to 25 percent soluble, and this includes enzymes. Carbohydrates, lipids and inorganic salts are particles in cytoplasm.

The outermost layer of cytoplasm, the plasmogel, can absorb water or remove it, and it is based on the cells need for liquid. This is called the stomatal guard cell in plants leaves.

The chemical composition of cytoplasm is 90 percent water and 10 percent of organic and inorganic compounds that vary in proportions.

Differences between Prokaryotic and Eukaryotic Cells

Prokaryotic cells belong to organisms such as bacteria, and they do not have a nucleus that is bound inside of the cells. In these types of cells, the cytoplasm is all of the contents of the cell that are bound by the outer cell membrane. In eukaryotic cells in plants, animals and humans, there is a nucleus, and the cytoplasm surrounding it has three main components of cytosol, organelles and cytoplasmic inclusions.

The nucleus of a cell is the command center. It is a structure containing the hereditary information, and its job is to control the growth and reproduction of a cell. The nucleus is the most prominent organelle in all cells. The nucleus is surrounded by a nuclear envelope which is a double membrane. It separates the contents of the nucleus from the cytoplasm with a double layer of lipids.

The envelope maintains the shape of the nucleus and regulates how the molecules flow both in and out of the nucleus through tiny holes called nuclear pores. The nucleus contains the chromosomes of DNA for heredity information and instructions that tell cells when to grow, develop and reproduce through chemical messages with other cells.

The cytosol is the liquid or semi fluid component in cytoplasm on the outside of the nucleus. Organelles perform specific functions in the cell. The cytoskeleton is located in the cytoplasm as fibers that help cells to maintain their shape, and they also provide support for organelles to survive and remain suspended in the liquid.

Organelles are tiny structures within a cell that each perform a specific function in the cell. Some examples of organelles are mitochondria, ribosomes, nucleus, lysosomes, chloroplasts, endoplasmic reticulum and Golgi apparatus.

Mitochondria generate power by the conversion of energy forms that the cell can use. The mitochondria are responsible for cellular respiration to generate fuel for the cells' activities from the food a person eats. You need to have energy at the cellular level to have cell division, cell growth and even cell death after division.

Ribosomes are organelles located in the cell that consist of proteins and your DNA. Ribosomes have the important and specific task of assembling all of the proteins in the cells. Ribosomes have a large and a small sub-unit that are synthesized in the nucleolus and then cross over to the cytoplasm through nuclear pores in the nuclear membrane. Ribosomes attach to messengers of RNA, and transfer it to the genetic material in proteins. They also link amino acids together, forming polypeptide chains that are modified and then become functional as proteins.

Lysosomes are sacs full of about 50 different enzymes that digest proteins, lipids and nucleic acids. It has a membrane to keep the internal compartment of the lysosome acidic, and it separates the digestive enzymes from the rest of the cell.

Chloroplasts are found in plant cells as an organelle. They store and collect substances that are needed for producing energy. It has a green pigment of chlorophyll to absorb light for photosynthesis, has its own DNA and reproduces in a process that is similar to binary fission of bacteria.

The endoplasmic reticulum plays the important role in producing, processing and transporting proteins and lipids for all the components in a cell.

The Golgi apparatus has the specific task of manufacturing, storing and shipping cellular products from the endoplasmic reticulum. There can be only a few Golgi apparatus or many in a cell depending on the type of cell.

Cytoplasmic inclusions are particles that are temporarily suspended in the cytoplasm of a cell. They may be macromolecular or granules such as secretory and nutritive inclusions and pigment granules. Secretory inclusions secrete something out of them such as acids, enzymes and proteins. Nutritive inclusions help give you nutrition such as the glucose storage molecules and lipids. The melanin in your skin cells is a pigment granule inclusion that controls your skin tone. Cytoplasmic inclusions are non-soluble and act as stored fats and sugars to use for cellular respiration.

Cyclosis

Cyclosis is also known as cytoplasmic streaming. It is the process by which substances move around in a cell. It occurs in different types of cells such as amoeba, fungi, plant cells and protozoa. The movement can be affected by temperature, light, chemicals or hormones.

Plants shuttle chloroplasts to areas that get the most sunlight, so they the plant organelles with the specific function of photosynthesis, which requires light. Amoeba and slime mold use this process for locomotion to move and capture food to survive. Cytoplasmic streaming is also required for both mitosis and meiosis in cell division to distribute the cytoplasm among the daughter cells from the parent cell.

Cyclosis occurs when the cytoplasm churns and creates a flow for materials through the cytosol. It can distribute nutrients and genetic information to pass through it from one organelle to the next organelle. For example, if one organelle produces a fatty acid or a steroid it can move through cyclosis to another organelle that needs it for good health in a cell. Cytoplasic streaming has another function of actually allowing a cell to move. In a cell with tiny hair like appendages outside the cell, the appendages allow them to move. In an amoeba the only manner in which a cell can move is through cyclosis.

Cytoplasm Working in Animal Cells

Animal cell cytoplasm is a gel-like material made of mostly water that fills the cells around the nucleus. It contains proteins and molecules that are particularly important for all cell health. The cytoplasm in an animal cell includes salts, sugars, amino acids, carbohydrates and nucleotides. Cytoplasm keeps all the cellular organelles suspended and helps in the movement of the cell through the cytoplasmic streaming process.

Cytoplasm Working in Plant Cells

Cytoplasm works in plant cells much like it does in animal cells. It provides support to the internal structures, is the suspension medium for the organelles and maintains the shape of a cell. It stores chemicals that are vital to plants for life and provides metabolic reactions such as synthesis of proteins and glycolysis. It supports cytoplasmic streaming around the vacuoles, which are spaces in the cytoplasm of a cell that are containing the fluid.

Cytoplasm Analogy

In order to see the large picture of a cytoplasm analogy of a restaurant it is best to represent the entire cell through an analogy.

The entire cell represents the entire restaurant, as it requires many different parts inside to function, just as cells have organelles for specific functions.

The cell membrane represents the restaurant doors as the restaurant doors allow people to enter and exit just as the membrane controls what items can enter and exit the entire cell.

The cytoplasm of a cell is represented by the restaurant floor. The restaurant floor holds tables, chairs and all objects in place, whereas the cytoplasm keeps all the organelles suspended in their places.

The nucleus of a cell is like a restaurant manager as the nucleus has control over what happens in the cell just as a restaurant manager controls the activities in the restaurant.

The cell's mitochondria are like the burger drawers to keep burgers warm until a customer orders their food. The mitochondria store all the energy that is obtained from the food and then share with the organelles when they need it.

The cell's endoplasmic reticulum is the same as the kitchen in the restaurant. The endoplasmic reticulum produces substances that are used in the cell and throughout the entire body such as fats and proteins that are needed for health. The kitchen produces many products that can be used in the restaurant, or they may be ordered at a drive through window for take out.

The cell's Golgi bodies and vesicles are akin to the front counter in a restaurant where employees put orders in bags to be eaten in the restaurant or in to go bags for customers to take with them to eat. The Golgi bodies serve to sort and transfer substances to be used in the cell or to transfer them out of the cell.

Plastids

Plastids are a group of phylogenetically and physiologically-related organelles found in all types of plants and algae. In their roles, the different types of plastids contribute to plant metabolism thus promoting plant growth and development. One of the main characteristics of these organelles is the fact that they have a double membrane.

In the cells, plastids are primarily involved in the manufacture and storage of food. They are therefore involved in such processes as photosynthesis, synthesis of amino acids and lipids as well as storage of various materials among a few other functions.

Apart from plants and algae, plastids can also be found in a number of other organisms including:

- Fern,

- Moss,

- Some parasitic worms,

- Some marine mollusks (some sea slugs).

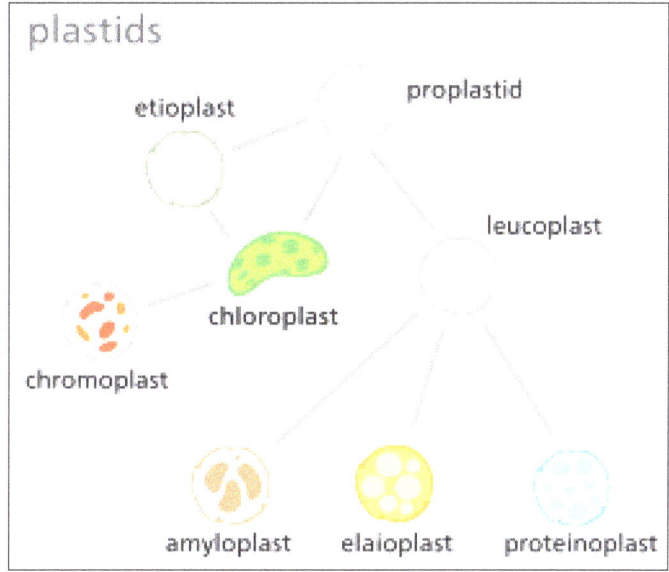

Types of Plastids.

Examples:

- Chromoplasts,

- Chloroplasts,

- Leucoplasts.

General Structure and Features of Plastids

For land plants, the number of plastids has been shown to be relatively high per cell ranging from 30 to 40 and 100 to 150 in diploid cells. The plastids of plants are also simpler when compared to those found in other organisms like algae.

Depending on the species (species of plant, algae etc) plastids may take up a variety of shapes ranging from discoid, spherical, dumbbell-shaped or lens-shaped among a few others.

Under stressful conditions, mitochondria have also been observed in plastids (by intrusion). This has been shown to be the case with plastids like chloroplast surrounding the organelle (mitochondria).

One of the other important structures associated with plastids is the stromule. By connecting the plastids into a network (plasidome) the stromule plays an important role in ensuring communication between the plastids and other cell organelles such as the mitochondria and the cell nucleus. Stromules are also highly dynamic and have been shown to extend from the surface of all types of plastids.

Apart from these aspects of plant plastids, some of the other features shared by all plastids include:

Double-membrane (Envelope Membrane)

For all types of plastids, the double membrane has been shown to be the only membrane that remains intact (permanent). It is made up of such galactolipids as MGDG among other lipids and proteins. Due to genome reduction of plastids, particularly in the cells, plastids are only able to encode for a small number of proteins.

As a result, they are highly dependent on the proteins encoded by the cell nucleus. Here, then, the double-membrane envelope of the plastids plays a crucial role in the transport of protein from the cell's cytoplasm and into the plastid.

Apart from protein transport, the membrane also plays an important role in the signaling process. Communication between plastids and the cell nucleus is important particularly during gene expression. The membrane, therefore, plays an important role in cell signaling and thus in the regulation of gene expression.

Some of the other roles of plastid envelopes include:

- Transport of other material including vital metals and metabolites.

- Metabolism of fatty acid, lipids, and carotenoid among other compounds.

- Production of plant growth regulators.

- Interaction with the cell's endomembrane systems.

Plastid Stoma

Stroma refers to the internal space that is enclosed by the double membrane of the plastid. It is filled by a colorless fluid/matrix that surrounds the thylakoid as well as a number of other organelles within the plastid.

Some of the other components of the stroma include:

Ribosome - is a major characteristic of plastid stroma. In some cells, they may be present as polyribosome, which is a complex of the mRNA molecule (a group of ribosome that are linked by the messenger RNA). In a plastid, the presence of ribosome indicates protein synthesis activities. Proteins are required for several functions including various chemical processes as well as damage repair. Therefore, the presence of a ribosome is essential for various plastid processes within a cell.

Nucleoids - These include copies of the plastid DNA and RNA. Like the cell nucleus, these nucleoids are the functional unit of the plastid's genome. Within the plastid, the nucleoids are attached to the thylakoids (in chloroplasts) or may be randomly spread in the stroma.

The number of nucleoids varies significantly from one organism to another. For instance, compared to non-green plastids, chloroplasts contain a higher number of nucleoids.

In plastids, the nucleoids may be organized along a ring and developed into a continuous ring of DNA. However, linear genomes have also been identified in plastids.

Like mitochondria, plastids are semi-autonomous bodies. As such, they contain their own genetic material and are therefore capable of synthesizing proteins required for normal functioning. However, close coordination between the plastids and the cell is important during plastid development given that they may depend on the cell for certain material required during processes.

Some of the other components of the plastid that may also be found in the stroma include:

- Inclusion bodies.
- Microtubules - E.g. etioplasts.
- Stromacenters.
- Starch.
- plastoglobuli.

Internal Membrane

The internal membrane of plastids is mostly found in land plants. It gradually develops from the inner membrane envelope (of the double membrane) as well as given lipid components.

In some cases, this membrane may attach to the inner membrane of the plastid to form a membrane system known as the peripheral reticulum. This system plays an important role in the transport of various materials from the cytoplasm of the cell and into the plastid and vice versa.

Chloroplast

Chloroplasts are organelles (compartments) found in plant cells and eukaryotic algae that conduct photosynthesis.

Utilizing chlorophyll and water, chloroplasts capture light energy from the sun to produce the free energy stored in ATP and NADPH through a process called photosynthesis. The photosynthetic mechanism also produces food for the organism in the form of sugar. In the process of creating sugar, carbon dioxide is consumed and oxygen is released. The simplified overall equation can be depicted as:

$$6\ CO_2 + 6H_2O + light \rightarrow C_6H_{12}O_6 + 6\ O_2$$

(Carbon Dioxide + Water + Light energy → Glucose + Oxygen)

As the source of all photosynthesis, the chloroplast serves a larger purpose by providing food for the entire plant or algae, and by extension for those organisms that consume these autotrophs.

The theorized origin of chloroplasts—developed through a mutually-beneficial symbiotic relationship between a cyanobacteria and a prokaryote—reflects a view, advocated by Lynn Margulis, that life developed more through cooperation than through competition.

Structure and Function

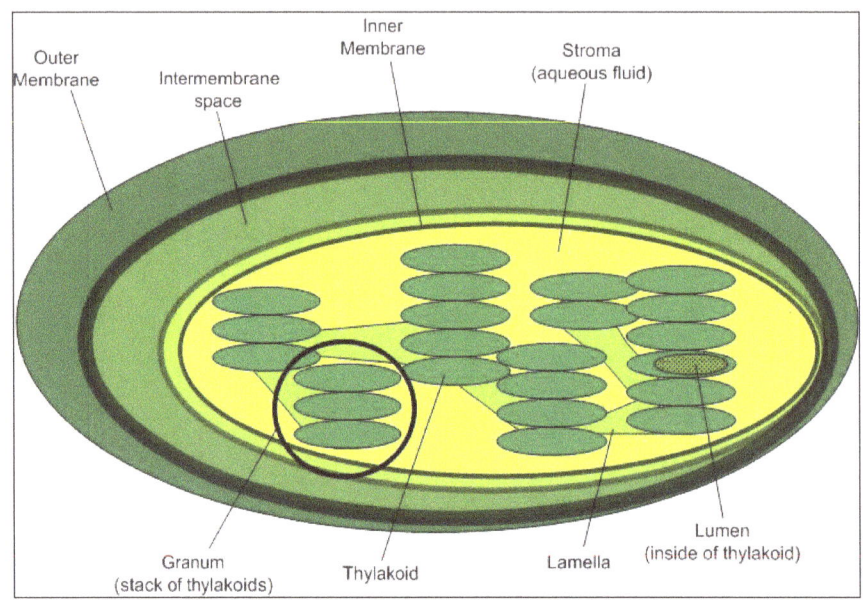

The inside of a chloroplast with a granum circled.

Chloroplasts are flat discs usually two to ten micrometers (μm) in diameter and one micrometer thick. The chloroplast has a two membrane envelope termed the inner membrane and outer membrane, respectively. Between these two layers is the intermembrane space.

The fluid within the chloroplast is called the stroma. Although most of the chloroplast's proteins are encoded by genes contained in the cell nucleus, some small portion of them are produced from tiny circular DNA and ribosomes present in the stroma, which corresponds to the cytoplasm of the bacterium.

Within the stroma also are stacks of thylakoids, the sub-organelles where photosynthesis actually takes place. A stack of thylakoids is called a granum (plural: grana). A thylakoid looks like a flattened disk with an empty inside region called the thylakoid space or lumen. The photosynthesis reaction takes place on the membrane of the thylakoid, and, as is also the case with mitochondria, utilizes the coupling of cross-membrane fluxes in a process to produce a new biological product. For photosynthesis, that product is both ATP and sugar.

Embedded in the thylakoid membrane is a dish of chlorophyll molecules known as an antenna complex. This outer array helps to increase the surface area of light capture. The light photons are then funneled to the center of this complex. Two chlorophyll molecules are then ionised, producing an excited electron, which then passes onto the photochemical reaction center.

The basic function of the chloroplast is photosynthesis. All photosynthetic reactions in the plant or algal cell occur in this organelle, and thus the chloroplast is the origin of all of the food (sugar) used by the other organelles of the plant or algae. Most of the sugars created in the chloroplast are converted by the plants into starch, which is stored in the plastids.

Origins

Chloroplasts are generally considered to have originated as endosymbiotic cyanobacteria. In this respect they are similar to mitochondria in assumed origin, but are found only in plants and protista. Both organelles are surrounded by a double celled composite membrane with an intermembrane space, both have their own DNA and are involved in energy metabolism, and both have reticulations, or many infoldings, filling their inner spaces.

In green plants, chloroplasts are surrounded by two lipid-bilayer membranes. The inner membrane is now thought to correspond to the outer membrane of the ancestral cyanobacterium. The chloroplast genome is considerably reduced compared to that of free-living cyanobacteria, but the parts that are still present show clear similarities. Many of the assumed missing genes are encoded in the nuclear genome of the host.

It is interesting to note that in some algae (such as the heterokonts and other protists such as Euglenozoa and Cercozoa), chloroplasts seem to have arisen through a secondary event of endosymbiosis, in which a eukaryotic cell engulfed a second eukaryotic cell containing chloroplasts, forming chloroplasts with three or four membrane layers. In some cases, such secondary endosymbionts are theorized to have themselves been engulfed by still other eukaryotes, forming tertiary endosymbionts.

Chromoplast

The coloration of the petals and sepals on the Bee orchid is controlled by a specialized organelle in plant cells called a chromoplast.

Chromoplasts are plastids, heterogeneous organelles responsible for pigment synthesis and storage in specific photosynthetic eukaryotes. It is thought that like all other plastids including chloroplasts and leucoplasts they are descended from symbiotic prokaryotes.

Function

Chromoplasts are found in fruits, flowers, roots, and stressed and aging leaves, and are responsible for their distinctive colors. This is always associated with a massive increase in the accumulation of carotenoid pigments. The conversion of chloroplasts to chromoplasts in ripening is a classic example.

They are generally found in mature tissues and are derived from preexisting mature plastids. Fruits and flowers are the most common structures for the biosynthesis of carotenoids, although other reactions occur there as well including the synthesis of sugars, starches, lipids, aromatic compounds, vitamins and hormones. The DNA in chloroplasts and chromoplasts is identical. One subtle difference in DNA was found after a liquid chromatography analysis of tomato chromoplasts was conducted, revealing increased cytosine methylation.

Chromoplasts synthesize and store pigments such as orange carotene, yellow xanthophylls, and various other red pigments. As such, their color varies depending on what pigment they contain. The main evolutionary purpose of chromoplasts is probably to attract pollinators or eaters of colored fruits, which help disperse seeds. However, they are also found in roots such as carrots and sweet potatoes. They allow the accumulation of large quantities of water-insoluble compounds in otherwise watery parts of plants.

When leaves change color in the autumn, it is due to the loss of green chlorophyll, which unmasks preexisting carotenoids. In this case, relatively little new carotenoid is produced—the change in plastid pigments associated with leaf senescence is somewhat different from the active conversion to chromoplasts observed in fruit and flowers.

There are some species of flowering plants that contain little to no carotenoids. In such cases there are plastids present within the petals that closely resemble chromoplasts and are sometimes visually indistinguishable. Anthocyanins and flavonoids located in the cell vacuoles are responsible for other colors of pigment.

The term "chromoplast" is occasionally used to include any plastid that has pigment, mostly to emphasize the difference between them and the various types of leucoplasts, plastids that have no pigments. In this sense, chloroplasts are a specific type of chromoplast. Still, "chromoplast" is more often used to denote plastids with pigments other than chlorophyll.

Structure and Classification

Using a light microscope chromoplasts can be differentiated and are classified into four main types. The first type is composed of proteic stroma with granules. The second is composed of protein crystals and amorphous pigment granules. The third type is composed of protein and pigment crystals. The fourth type is a chromoplast which only contains crystals. An electron microscope reveals even more, allowing for the identification of substructures such as globules, crystals, membranes, fibrils and tubules. The substructures found in chromoplasts are not found in the mature plastid that it divided from.

The presence, frequency and identification of substructures using an electron microscope has led to further classification, dividing chromoplasts into five main categories: Globular chromoplasts, crystalline chromoplasts, fibrillar chromoplasts, tubular chromoplasts and membranous chromoplasts. It has also been found that different types of chromoplasts can coexist in the same organ. Some examples of plants in the various categories include mangoes, which have globular chromoplasts, and carrots which have crystalline chromoplasts.

Although some chromoplasts are easily categorized, others have characteristics from multiple categories that make them hard to place. Tomatoes accumulate carotenoids, mainly lycopene crystalloids in membrane-shaped structures, which could place them in either the crystalline or membranous category.

Leucoplast

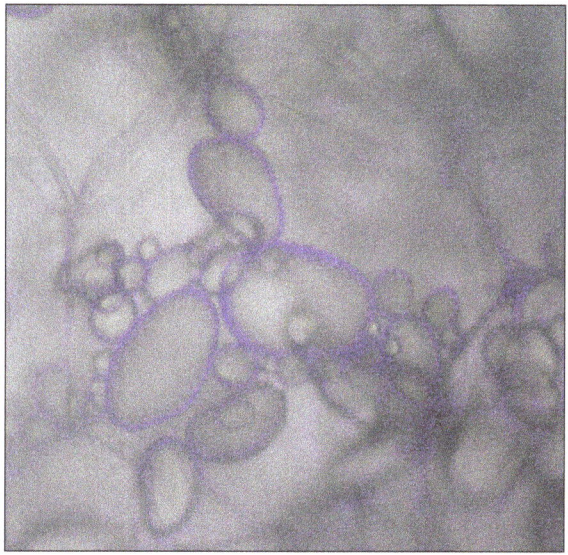

Leucoplasts, specifically, amyloplasts.

Leucoplasts are a category of plastid and as such are organelles found in plant cells. They are non-pigmented, in contrast to other plastids such as the chloroplast.

Lacking photosynthetic pigments, leucoplasts are not green and are located in non-photosynthetic tissues of plants, such as roots, bulbs and seeds. They may be specialized for bulk storage of starch, lipid or protein and are then known as amyloplasts, elaioplasts, or proteinoplasts (also called aleuroplasts) respectively. However, in many cell types, leucoplasts do not have a major storage function and are present to provide a wide range of essential biosynthetic functions, including the synthesis of fatty acids such as palmitic acid, many amino acids, and tetrapyrrole compounds such as heme. In general, leucoplasts are much smaller than chloroplasts and have a variable morphology, often described as amoeboid. Extensive networks of stromules interconnecting leucoplasts have been observed in epidermal cells of roots, hypocotyls, and petals, and in callus and suspension culture cells of tobacco. In some cell types at certain stages of development, leucoplasts are clustered around the nucleus with stromules extending to the cell periphery, as observed for proplastids in the root meristem.

Etioplasts, which are pre-granal, immature chloroplasts but can also be chloroplasts that have been deprived of light, lack active pigment and can be considered leucoplasts. After several minutes exposure to light, etioplasts begin to transform into functioning chloroplasts and cease being leucoplasts. Amyloplasts are of large size and store starch. Proteinoplasts store proteins and are found in seeds (pulses). Elaioplasts store fats and oils and are found in seeds. They are also called oleosomes. [castor, groundnut] Etioplasts are plastids without pigments and store food and lamellar structures. These plastids occur in etiolated plants due to the absence of light.

Elaioplast

An elaioplast is a leucoplast that is primarily involved in storing fats or lipids inside fat droplets (plastoglobuli) in plants (particularly in monocots and liverworts). Plastoglobuli are spherical bubbles containing lipids such as steryl esters. Nevertheless, plastoglobuli are not exclusive to elaioplast. They also occur in other plastids such as chloroplasts particularly when the latter are under oxidative stress or would undergo transformation into gerontoplast.

Elaioplasts are most intensively studied in tapetal cells where they play an essential role in pollen maturation. Tapetal cells have elaioplasts and tapetosomes (oil and protein bodies derived from the endoplasmic reticulum). Both the elaioplasts and tapetosomes contribute to the formation of pollen coat during the final stage of pollen maturation. The elaioplast, in particular, are released by the tapetal cell through lysis. The sterol lipids of the elaioplast coat the outside of the pollen grain. The tapetosomes, in turn, provide proteins to the pollen coat. Elaioplasts should not be confused with oleosomes, which are derived from rough endoplasmic reticulum and stores oil as well. The oleosomes are found primarily in seeds. They are probably used mainly for longer-term oil storage compared with the elaioplasts that are for shorter-term oil storage and synthesis.

Proteinoplast

Proteinoplasts are leucoplasts that contain crystalline bodies of proteins. They may also serve as a site for certain enzymatic activities. However, the proteinoplasts contain protein inclusions that may be crystalline or amorphous and often enclosed by a membrane. Proteinoplasts are found in seeds, e.g. brazil nuts, peanuts, etc. Compared with chloroplast, which is a green plastid involved in photosynthesis, the proteinoplast has fewer thylakoids.

Amyloplast

An amyloplast is an organelle found in plant cells. Amyloplasts are plastids that produce and store starch within internal membrane compartments. They are commonly found in vegetative plant tissues, such as tubers (potatoes) and bulbs. Amyloplasts are also thought to be involved in gravity sensing (gravitropism) and helping plant roots grow in a downward direction.

Amyloplasts are derived from a group of plastids known as leucoplasts. Leucoplasts have no pigmentation and appear colorless. Several other types of plastids are found within plant cells including chloroplasts (sites of photosynthesis), chromoplasts (produce plant pigments), and gerontoplasts (degraded chloroplasts).

Amyloplast Development

Amyloplasts are responsible for all starch synthesis in plants. They are found in plant parenchyma tissue which composes the outer and inner layers of stems and roots; the middle layer of leaves; and the soft tissue in fruits. Amyloplasts develop from proplastids and divide by the process of binary fission. Maturing amyloplasts develop internal membranes which create compartments for the storage of starch.

Starch is a polymer of glucose that exists in two forms: amylopectin and amylose. Starch granules are composed of both amylopectin and amylose molecules arranged in a highly organized fashion. The size and number of starch grains contained within amyloplasts varies based on the plant species. Some contain a single spherical shaped grain, while others contain multiple small grains. The size of the amyloplast itself depends on the amount of starch being stored.

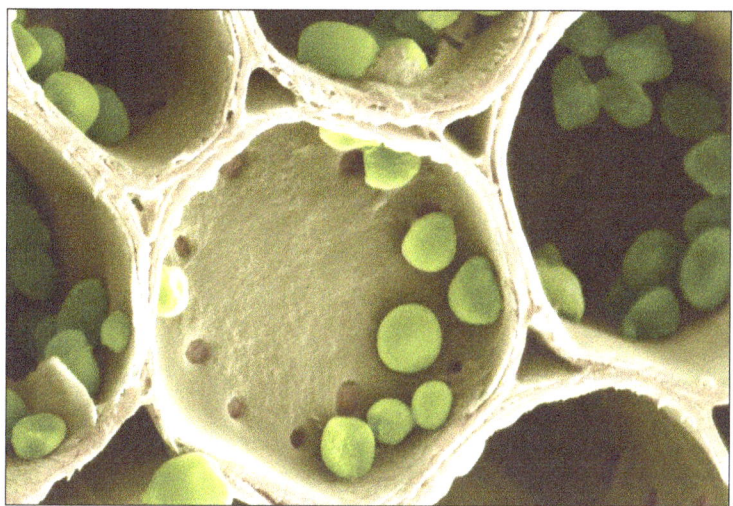

This image shows starch grains (green) in the parenchyma of a Clematis sp. plant. Starch is synthesized from the carbohydrate sucrose, a sugar produced by the plant during photosynthesis, and used as a source of energy. It is stored as grains in structures called amyloplasts (yellow).

Tannosome

Tannosomes are organelles found in plant cells of vascular plants.

Formation and Functions

Tannosomes are formed when the chloroplast membrane forms pockets filled with tannin. Slowly, the pockets break off as tiny vacuoles that carry tannins to the large vacuole filled with acidic fluid. Tannins are then released into the vacuole and stored inside as tannin accretions.

They are responsible for synthesizing and producing condensed tannins and polyphenols. Tannosomes condense tannins in chlorophyllous organs, providing defenses against herbivores and pathogens, and protection against UV radiation.

The Photosynthetic Machinery

The thylakoid membrane houses chlorophylls and different protein complexes, including photosystem I, photosystem II, and ATP (adenosine triphosphate) synthase, which are specialized for light-dependent photosynthesis. When sunlight strikes the thylakoids, the light energy excites chlorophyll pigments, causing them to give up electrons. The electrons then enter the electron transport chain, a series of reactions that ultimately drives the phosphorylation of adenosine diphosphate (ADP) to the energy-rich storage compound ATP. Electron transport also results in the production of the reducing agent nicotinamide adenine dinucleotide phosphate (NADPH).

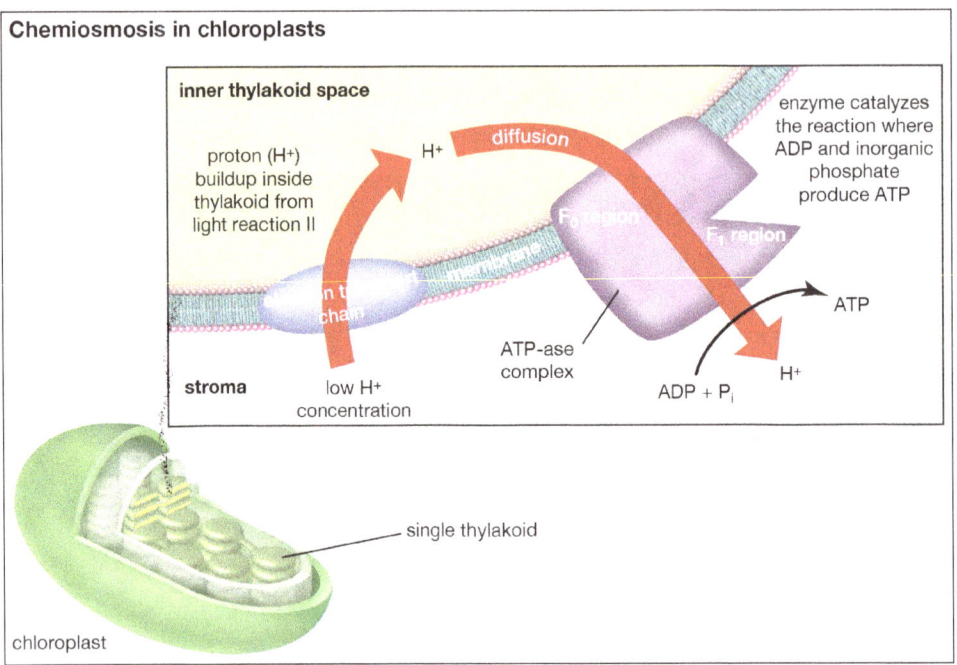

chemiosmosis in chloroplasts Chemiosmosis in chloroplasts that results in the donation of a proton for the production of adenosine triphosphate (ATP) in plants.

ATP and NADPH are used in the light-independent reactions (dark reactions) of photosynthesis, in which carbon dioxide and water are assimilated into organic compounds. The light-independent reactions of photosynthesis are carried out in the chloroplast stroma, which contains the enzyme ribulose-1,5-bisphosphate carboxylase/oxygenase (rubisco). Rubisco catalyzes the first step of carbon fixation in the Calvin cycle (also called Calvin-Benson cycle), the primary pathway of carbon transport in plants. Among so-called C_4 plants, the initial carbon fixation step and the Calvin cycle are separated spatially—carbon fixation occurs via phosphoenolpyruvate (PEP) carboxylation in chloroplasts located in the mesophyll, while malate, the four-carbon product of that process, is transported to chloroplasts in bundle-sheath cells, where the Calvin cycle is carried out. C_4 photosynthesis attempts to minimize the loss of carbon dioxide to photorespiration. In plants that use crassulacean acid metabolism (CAM), PEP carboxylation and the Calvin cycle are separated temporally in chloroplasts, the former taking place at night and the latter during the day. The CAM pathway allows plants to carry out photosynthesis with minimal water loss.

Cell Plate

The cell plate is a structure that forms in the cells of land plants while they are undergoing cell division.

The cells of land plants, unlike animal cells, have a cell wall made of stiff sugars which surround their cell membranes. In addition to protecting the cell from damage, the cell walls help to maintain the plant's rigid upright structures, such as leaves and stems.

These rigid support structures allow plants to grow tall and spread their leaves wide, obtaining more sunlight. In most plants, the cell wall is made of cellulose – an arrangement of glucose molecules that forms hard, rigid surfaces.

Interestingly, the cellulose that makes up cell walls is not digestible to humans or animals – but it can be broken down into sugar by some methane-producing archaebacteria. This is one reason for the symbiotic relationship between many animals and the archaebacteria in our gut.

During cell division, plant cells must form a new cell wall to separate their daughter cells. This new fragment of cell wall must form in the middle of the parent cell, to ensure that half of the parent cells' chloroplasts, gene copies, etc. end up on each side of the cell wall.

The "plate" of hard sugars that forms in the middle of the parent cell, which will become the cell wall of the future daughter cells, is called the cell plate.

It is formed when vesicles from the Golgi apparatus carrying phospholipids needed to make the cell membrane, and sugars needed to form the cell wall, are delivered and assembled along a network of cytoskeleton spindle fibers that forms in the middle of the cell as the cell prepares to divide.

Function of Cell Plate

Cell walls serve the double purpose of protecting the precious contents of plant cells, such as their nuclei, and allowing a plant to have free-standing structure.

Since plants do not have skeletons like animals, and are constantly growing and changing in competition to get more sunlight, it is important that the individual parts of plants, such as stems and leaves, be able to stand straight against the force of gravity on their own.

This is why land plants have cell walls, but animals who have skeletons, and ocean plants that live in the weightless environment under water may not have cell walls.

Having cell walls makes cell division a bit tricky for plants. To split in two and produce daughter cells – a process called "cytokinesis" – cells without cell walls simply pinch their cell membrane in two around the middle. The cell membrane is like a flexible bag, which can be pinched and re-formed as needed when a cell needs to change shape.

The rigid cell wall, however, cannot be bent or pinched in the same way. It constrains the shape of a cell during reproduction, and cannot be simply pinched around the middle.

Instead, to perform cytokinesis, plants must assemble a new section of cell wall to ensure that their daughter cells will have the structural integrity that the plant needs to maintain its form.

Cell Plate Formation

The "cell cycle" describes the process that cells go through, from their "birth" as new daughter cells, until they themselves are ready to split and become "parent cells" to two new daughter cells.

The formation of the cell plate takes place during the mitotic phase. In this description we will briefly describe all phases of the cell cycle to paint a complete picture, but feel free to skip to the section labeled "mitotic phase" to get a play-by-play on how the cell plate forms.

The stages of the cell cycle are divided into:

- Interphase – where the cell grows and matures.

- The mitotic phase – where the cell begins working towards dividing itself.

Interphase

These stages are further developed into specific steps in which the cell takes all the actions needed to produce two healthy daughter cells. The steps of interphase are:

- G1 Phase – Immediately after a daughter cell becomes independent, it spends some time growing and creating more organelles, such as chloroplasts.

- S Phase – In S Phase – shirt for "Synthesis" phase – the cell synthesizes a new copy of its DNA. All of its chromosomes are copied, as is its centrosome, which will help ensure that one copy of each chromosome goes to each daughter cell during cell division.

- G2 Phase – The cell continues to make more proteins and materials for its daughter cells, and cytoskeleton structures begin to form which will help the cell perform cytokinesis – and, in the case of land plant cells, form their cell plate.

Mitotic Phase

After G2 comes the mitotic phase, where the cell begins taking the actions it needs to split.

For the cell plate to form, space must be cleared of any vacuoles or any other obstructions that could get in the way; the chromosomes must be assorted so that each daughter cell receives a copy of each; and then the cell plate can form, separating the cytoplasm of the two daughter cells.

All that, and more, occurs during the mitotic phase.

The stages of the mitotic phase are:

- Preprophase. Some plant cells have extremely large vacuoles, or vesicles that store materials such as fuel. In plant cells like these, the vacuoles first need to be moved or split in half to create a clear path through the cell's cytoplasm where the cell plate can form. The nucleus also needs to be moved into the center of this path, so that its DNA can be split between the two daughter cells. This clear band of cytoplasm is called the phragmosome.

- Prophase, in which the cell's DNA curls up tightly into compact chromosomes. These compact chromosomes are easier to sort and move, ensuring that no DNA is left out of either daughter cell. The mitotic spindle, which will help move chromosomes to opposite sides of the cell which will then become separate daughter cells, also begins to form. In land plants, the mitotic spindle also lays the groundwork for the cell plate to form along the phragmosome.

- In metaphase, motor proteins line the chromosomes up along a network of microtubules in the middle of the cell. This network is called the "metaphase plate." In plant that have a Phragmosome, the metaphase plate usually runs through the Phragmosome, and occupies the same location where the cell plate will later form.

- In anaphase, chromosomes lined up along the metaphase plate, chromosomes are separated, with one copy of each being dragged away from the center and toward each end of the cell. In this way, one copy of each chromosome ends up in the cytoplasm of what will become the daughter cells.

- In telophase, a new nucleus forms in each daughter cell. This happens in the daughter cells' cytoplasm, away from the cell plate.

- The final phase of cell division, following telophase, is cytokinesis. In this stage the cell's cytoplasm splits in two to produce two daughter cells. This is when the cell plate is formed. First, a network of microtubules, microfilaments, and parts of the parent cell's endoplasmic reticulum form a structure called the "phragmoplast."With the phragmoplast serving as its scaffold, the cell plate begins to assemble. Vesicles from the Golgi apparatus bring phospholipids to form the daughter cells' new cell membranes, and sugars to form their rigid cell walls. The growth of the cell plate along the phragmoplast is pictured below:

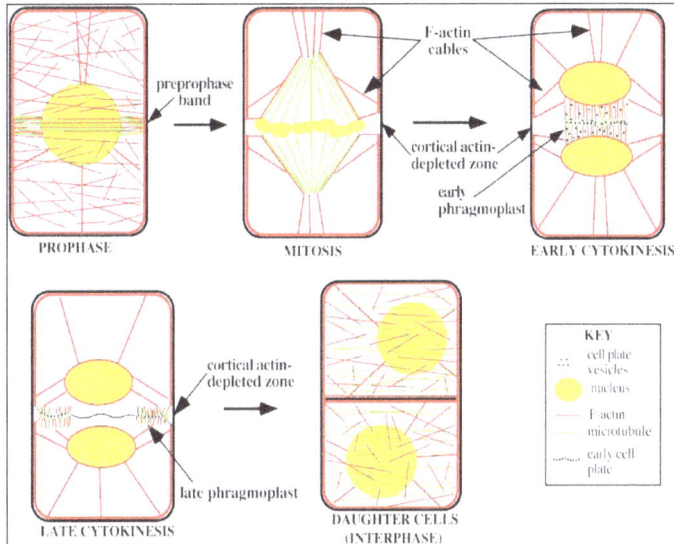

The process of cell division for plants is truly complete when the cell plate has fused with the cell walls, forming a fully functioning divider between the two daughter cells.

Cytokinesis

Cytokinesis is the partitioning of the cytoplasm following nuclear division. This process presents a number of challenges for the plant cell: first, to avoid losing or bisecting the nucleus, this event needs to be carefully coordinated with respect to the nuclear cycle in space and in time. Second, a structure as complex as the plant cell wall needs to be laid down during the brief period of time between anaphase and telophase. The spatial and temporal regulation of cytokinesis requires a series of links between the nuclear cycle, the cell cortex, the Golgi apparatus, and the membrane trafficking apparatus. A number of genes have recently been identified that could enable us to probe such links.

Cytokinesis in higher plants may be considered as a specialized form of secretion. At the end of anaphase, Golgi-derived secretory vesicles carrying cell wall materials are transported to the equator of a dividing cell. Fusion of these vesicles gives rise to a membrane-bound compartment, the cell plate. The cell plate expands from the middle out (centrifugally) until it reaches the "zone of attachment" or division site on the mother cell wall. Once this attachment has taken place, the cell plate undergoes a complex process of maturation during which callose is replaced by cellulose and pectin.

Two cytoskeletal arrays, the preprophase band (PPB) and the phragmoplast, play central roles in cytokinesis in the somatic cells of higher plants. The PPB, a transient ring of cortical microtubules (MTs) and actin filaments, appears in late S phase, narrows throughout G2, and disappears during prophase when the nuclear envelope (NE) breaks down. The phragmoplast is an array of MTs and actin filaments present at the equator of a dividing cell during the anaphase to telophase transition.

The master choreographer of cytokinesis is the nucleus. Phragmoplast MTs are thought to be remnants of the mitotic spindle, which acts as a scaffold to ensure continuity in space and time between the nucleus and the cell equator. A number of processes underlying cytokinesis, including cytoskeletal reorganization, the biosynthesis and packaging of cell wall polymers, and vesicle traffic, need to be tightly regulated with respect to the nuclear cycle. It is possible that some of these processes are regulated by the numerous kinases found at the phragmoplast.

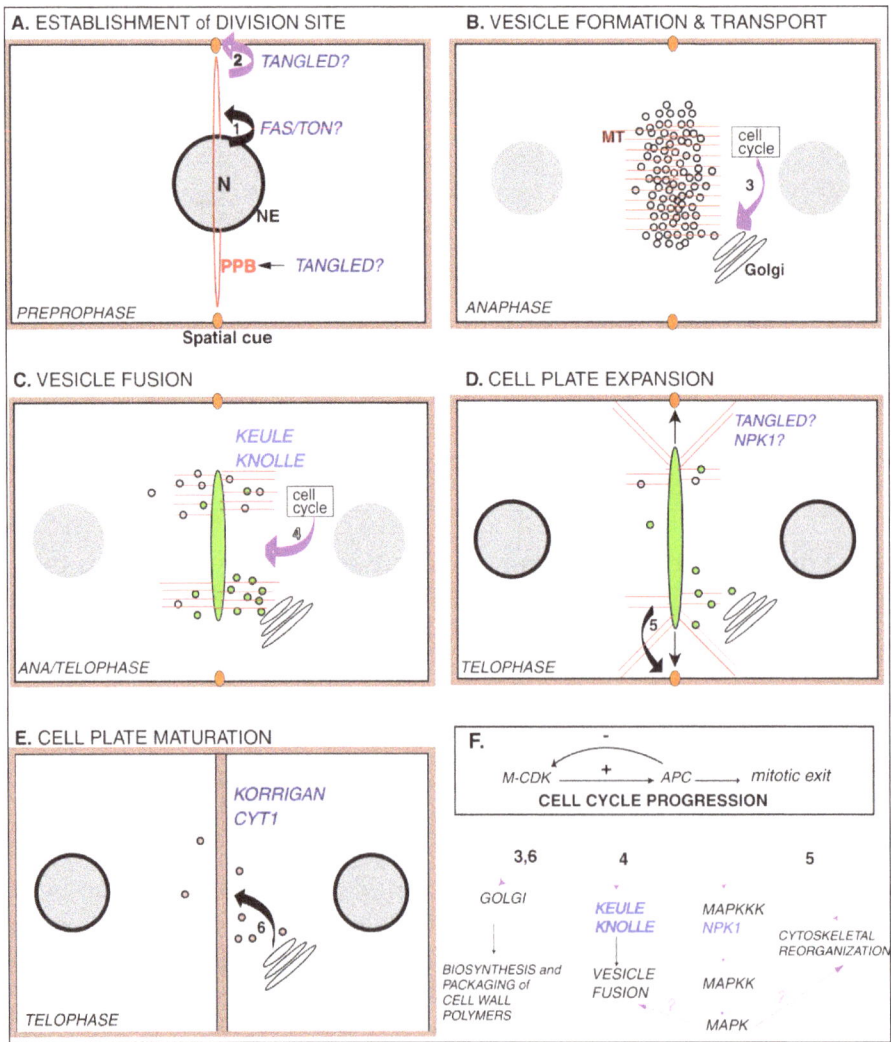

Plant cytokinesis: exploring the links. A, The division site is established during preprophase. The nuclear surface organizes the PPB (1), which may establish spatial cues (orange ovals; 2). B, Cell wall polymers such as xyloglucans are synthesized at the Golgi, packaged into vesicles, and transported to the equator of a dividing cell. C, The cell plate arises via fusion of Golgi-derived vesicles at the equator. The plate expands through the addition of secretory vesicles at the periphery, the deposition of callose in the lumena of the vesicular network, and via the concentric displacement of MTs toward the cell cortex (compare MTs in B, C, and D). Callose synthesis occurs within the cell plate, and this may be triggered by high concentrations of membrane-associated calcium. D, The phragmoplast is guided toward the spatial cue established during preprophase. MTs are now reoriented, with their minus ends at the equator and their plus ends toward the cell cortex. E, The cell plate fuses with the parental membranes and cell wall at the division site. Cellulose biosynthesis occurs at the plasma membrane; pectin biosynthesis takes place within the Golgi. F, The biosynthesis and/or packaging of cell wall polymers, membrane traffic, a MAP kinase cascade, and cytoskeletal organization are presumably coordinated with respect to cell cycle progression. How this relates to anaphase-promoting complex (APC) activation is unclear. Also, the targets of the MAP kinase remain to be determined. PPB and phragmoplast MTs are shown in red. Black arrows highlight established links and purple arrows potential links. Genes required for plant cytokinesis are shown in blue. The NE breaks down during mitosis and reassembles during telophase. The immature cell plate, rich in callose and also containing xyloglucans, is colored green, and the mature walls containing cellulose and pectin are colored in brown.

Spatial Control of Cytokinesis

Elegant experiments involving the displacement of nuclei or immature cell plates by centrifugation in moss protonemata or stamen hair cells have clearly demonstrated that the position of the interphase nucleus determines the division plane. Within a narrow window of time during the cell cycle, the nucleus is competent to dictate the position of the PPB that, in turn, marks the future division site. In stamen hair cells, immature cell plates displaced by centrifugation curve toward the site formerly occupied by the PPB. Thus, a spatial cue is laid down early during the cell cycle and, later during the cell cycle, the nascent cell plate is guided toward this cue.

A relay of links between the nucleus, the PPB, and the cell cortex/plasma membrane/cell wall may be implicated. First, the surfaces of plant nuclei have been shown to possess all the properties of MT organizing centers (MTOCs). These properties include: (a) the ability to nucleate and organize MTs, (b) the capacity to establish and/or anchor the minus ends of MTs, and (c) the presence at the nuclear surface of gamma tubulin and other proteins characteristic of animal MTOCs (Stoppin et al., 1994). Thus, it is likely that perinuclear MTOCs, which are associated with the NE, organize the PPB during late S-phase.

The division site plays an important role in cell wall maturation. Due to their high content in callose, immature cell plates are fluid and wrinkled. In contrast, mature cell plates are stiff and flat. This change occurs after the expanding cell plate has reached the division site and is accompanied by callose removal and cellulose and pectin deposition. In 1990, Mineyuki and Gunning proposed that the division site is established by: "(1) localized deposition of insertion and maturation factors in a latent form; and (2) provision of a means that, later on, will guide the leading edge of the centrifugally extending phragmoplast to the site. Once the new wall has attached, the factors are activated and utilized to insert, anchor and integrate new wall and contribute centripetally to its development. The PPB's raison d'être is to provide the necessary spatial guidance for the localized deposition of the(se) factors." Evidence for this model includes the following observations: (a) Cell plate maturation occurs if the nascent cell plate is inserted at the division site, but not if it is caused to insert elsewhere; (b) cell wall stubs develop from the outside in (centripetally) in cytokinesis-defective mutants or in caffeine-treated cells; (c) visible alterations of the parental cell wall occur at the site underlying the PPB; (d) PPBs do not occur in cells types in which the new walls are not inserted into parental walls; and (e) the position of the PPB and of the division site invariably correlate.

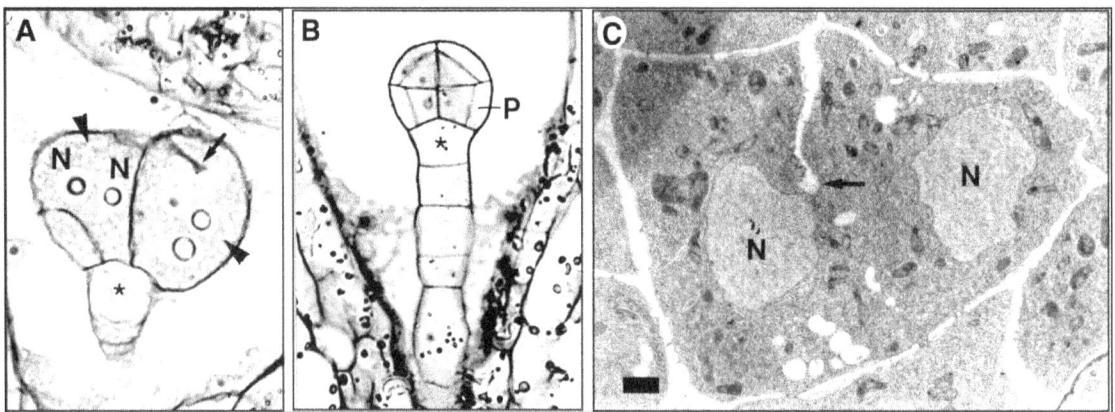

Cell wall stubs and mutlinucleate cells in cytokinesis-defective embryos. A and B, Light micrographs of dermatogen stage embryos. A, keule mutant. B, Wild type. C, Electron micrograph of keule mutant embryo at dermatogen stage showing a non-vacuolated binuleate cell. The centripetal cell wall stub supports the model that the division site consolidates the immature cell plate. N, Nucleus; P, protoderm. A star marks the uppermost cell of the suspensor. Arrows point to cell wall stubs and arrowheads to multinucleate cells.

Genes Required for Cell Wall Orientation

The above model is strengthened by the analysis of plant mutants impaired in their ability to orient cell walls. The Arabidopsisfas and tonneau mutants, as well as thetangled 1 mutants of maize (Zea mays), are characterized by misoriented cell walls, especially during asymmetric or longitudinal divisions; whereas fas andtonneau mutants altogether lack PPBs, these rings of cortical MTs are often misoriented in tangled mutants. In tonneau mutants, cortical MTs in general, including the PPB, are perturbed, yet the spindle and phragmoplast appear normal. It is possible that the FAS/TONNEAU gene products function at the nuclear surface to organize cortical MT arrays.

The TANGLED1 gene has been cloned and characterized. The gene encodes a highly basic protein bearing little sequence similarity to other proteins, yet possessing domains weakly homologous to the MT binding domain of vertebrate APC. TAN1 binds to MTs in vitro, possibly in a cell cycle-dependent manner, and proteins recognized by anti-TAN1 antibodies localize to the PPB, spindle, and phragmoplast in dividing cells, providing evidence thatTANGLED1 may encode an MT-binding protein. It is interesting that in tangled mutants the leading edges of phragmoplasts are not guided to sites formerly occupied by PPBs. Mutant cell plates consistently do not undergo the flattening that accompanies cell plate maturation, but remain wrinkled. In addition to orienting the PPB, TANGLED may be implicated in the establishment of the division site during preprophase and/or may guide the leading edges of the phragmoplast to this site during cytokinesis. The molecular identity of the spatial cue that determines the division site, as well as its localization to the cortex, plasma membrane, and/or cell wall, remain to be determined.

Temporal Regulation of Cytokinesis

Cell Cycle Progression: Exiting Mitosis and Initiating Cytokinesis

The onset of cytokinesis is concomitant with exit from mitosis. In plants, progression through mitosis relies on the activity of the cyclinB-cdc2 complex (M-CDK) active during M phase. Shortly before anaphase, M-CDKs are thought to activate the APC, a ubiquitin ligase that in turn destroys the M-CDKs. Thus, cyclin-dependent kinases and the APC regulate each other to ensure timely progression through mitosis. A number of studies in budding and fission yeast (Saccharomyces cerevisiae and Saccharomyces pombe) support the notion that mitotic exit does not alone suffice for the initiation of cytokinesis. In fact, an additional kinase cascade triggers a cytokinetic pathway. In budding yeast, the polo kinase CDC5 participates in APC activation and, in addition, appears to regulate the cytokinetic pathway by interacting with septins. Because plants lack septins, it is not clear how these findings in yeast relate to plant cytokinesis.

A large number of kinases have been found at the phragmoplast and these are good candidates for orchestrating the onset and execution of cytokinesis. Compelling evidence that a MAP kinase cascade is required for plant cytokinesis comes from a recent study showing that kinase negative mutations inNPK1, a MAP kinase kinase kinase, disrupt cytokinesis in tobacco (Nicotiana tabacum) cells. NPK1 activity is up-regulated during late M phase. The protein is present in the nucleus during interphase, and at the equatorial zone of the phragmoplast where it may be required for phragmoplast expansion toward the cell cortex. Thus, NPK1 may provide continuity in space and in time between the interphase nucleus and the cell equator. A MAP kinase has been detected at the phragmoplast in alfalfa (Medicago sativa) cells and its tobacco orthologue might be a target of NPK1. Two important questions remain unanswered: First, what signals activate NPK1? And, second, what are the targets of the putative MAP kinase cascade downstream of NPK1?

Cytokinesis and the Nuclear Cycle: One-way Communication

The multinucleate phenotype of cytokinesis-defective mutants suggest that nuclear division can be initiated and completed even if cytokinesis is incomplete. In contrast, cell wall stubs in such mutants are only observed in multinucleate cells, which suggests that cytokinesis can only be initiated once the nuclear cycle is complete. Titan and pilz mutants, characterized by giant nuclei, consistently show marked cytokinesis defects. A simple hypothesis is that the cytokinesis defects observed in titan andpilz mutants are an indirect consequence of cell cycle arrest due to a primary defect in nuclear division. This is supported by the observation, based on tubulin stains, that not only cytokinesis but cell cycle progression in general is affected in pilzmutants. A number of conserved cell cycle checkpoints known to monitor nuclear division and spindle assembly or orientation readily account for these observations. In this context, it is interesting to note that cytokinesis defective mutants such as keule have enlarged nuclei, suggesting that incomplete cytokinesis may also impact the nuclear cycle, though not in an absolute way. In general, however, the nucleus appears to play a dominant role in dictating the onset of cytokinesis.

Plant Cytokinesis as a Specialized Form of Secretion

Role of the Golgi Apparatus in Cytokinesis

Drug studies have highlighted the importance of the Golgi apparatus in cytokinesis. The biosynthesis and assembly of numerous cell wall polysaccharides take place in the Golgi. In contrast, callose is synthesized within the cell plate and cellulose microfibrils are synthesized by cellulose synthases embedded in the plasma membrane, which explains why cell plate flattening and maturation only occur after fusion with the parental membrane and wall. Immature cell plates are rich in xyloglucans and devoid of pectins, whereas mature cross walls are rich in pectins and have low xyloglucan content. Because pectins and xyloglucans are synthesized in the Golgi and targeted to the cell plate, there appears to be a tight cell cycle regulation of Golgi activity and secretion.

The Membrane Dynamics of Cytokinesis

At the end of anaphase, Golgi-derived vesicles are transported toward the equator of a dividing cell. Vesicle fusion gives rise to the cell plate, assembled within the phragmoplast. These vesicle trafficking events can be broken down into four steps: vesicle formation, transport, tethering/docking, and fusion. As shown in figure, each of these steps is highly regulated.

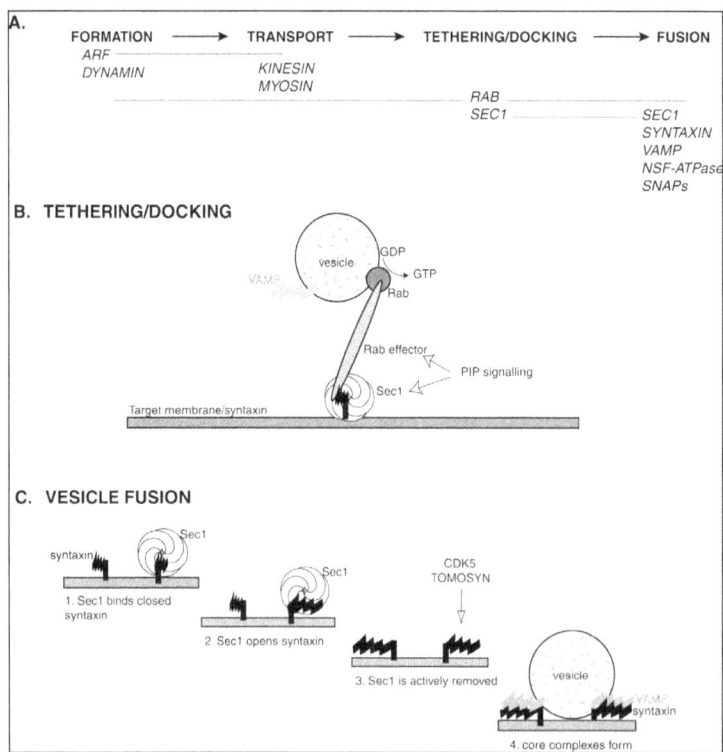

A, Vesicle trafficking can be broken down into four steps: formation, transport, tethering/docking, and fusion. Some of the key players required at each step are indicated in italics, and lines denote their range of action. ADP-ribosylation factor (ARF) function is best documented for vesicle formation, and Rabs regulate vesicle docking. Through a variety of effectors and through potential cross talk, both ARF and Rab GTPases in principle could regulate all four steps, as is thought to be the case for Rabs. B, Vesicle docking is regulated by Rabs and Sec1s through the activity of rab effectors. These are large complexes or long molecules capable of multiple interactions with the Rab on the vesicle and the Sec1 and syntaxin on target membranes. These interactions may be regulated by phosphatidylinositol (PIP) signaling. C, Vesicle fusion requires an interaction between vesicle-associated membrane proteins (VAMPs) on vesicles and syntaxins on target membranes; based on studies of neural cells and synapses, it is thought that syntaxins need to be primed or opened for this interaction to take place. Sec1 proteins bind the closed syntaxin (1) and induce a conformational change (2). The Sec1 is then actively removed (3) with the help of CDK5 and tomosyn. 4, With the help of SNAP25-like adapters, a core complex is subsequently formed between the VAMP on the vesicle and the syntaxin on the target membrane. This pulls the membranes close together, overcoming the barriers to fusion. For the sake of simplicity, we present models based on neural systems for which the vesicle and target membranes are distinct. As regards the initial rounds of fusion at the cell plate, however, these membranes are in all likelihood identical.

Vesicle Formation

An interesting recent finding is that one of the TITANgenes, TITAN5, encodes an ARF GTPase that may be implicated in vesicle formation during cell division. This ARF may, for example, be required for NE breakdown into vesicles, and this could explain the giant nuclei observed in titan5 mutant embryos and endosperm. In addition to a role in regulating nuclear division, TITAN5 could also be required for vesicle formation during cytokinesis. With 45 members, ARFs represent a large gene family in Arabidopsis, and therefore it is likely that individual ARFs have specialized as opposed to multiple functions. Because plants lack orthologues of nuclear lamins, the identification of TITAN 5 provides a highly important and novel handle on the poorly understood process of NE dynamics and function.

Vesicle Transport

At anaphase, vesicles are transported to the equator of a dividing cell. Because the plus ends of phragmoplast MTs overlap at the equator, a plus end-directed motor such as kinesin would transport Golgi-derived vesicles to the equatorial region during cytokinesis. In addition, a minus end-directed motor such as myosin may play a role in vesicle translocation to the phragmoplast. Both kinesin and myosin have been found at the phragmoplast.

Vesicle Tethering or Docking

Prior to fusion, vesicles are found docked at their target membranes. Docking is mediated by a number of protein interactions that connect Rab GTPases on vesicles with syntaxins on target membranes. Although Arabidopsis contains a large family of Rab proteins, none has been implicated thus far in cytokinesis.

Vesicle Fusion

Vesicle fusion requires a specific "lock and key" interaction between syntaxins on target membranes and VAMPs or v-SNAREs on vesicle membranes. Members of the Sec1 superfamily of proteins appear capable of inducing conformational changes in syntaxins, "opening" or priming these for interactions with other SNAREs. Molecular genetic analyses have identified two genes whose products, KEULE and KNOLLE, concertedly mediate membrane fusion events during cytokinesis. At a cellular level, knolle and keule mutants are characterized by multinucleate cells with gapped or incomplete cross walls. KNOLLE encodes a cytokinesis-specific syntaxin expressed in punctate, vesicle-like structures during M phase, and at the phragmoplast in dividing cells. KEULE encodes a Sec1 protein that has been shown to bind KNOLLE in in vitro-binding assays. The observation that vesicles accumulate but do not fuse at the equator of dividing cells in keule and knolle embryos indicates that, like the syntaxin KNOLLE, KEULE is required for vesicle fusion. The synthetic lethality of knolle keule double mutants has provided evidence that the two genes interact in vivo. Whereas cytokinesis is impaired but not blocked in keule and knolle mutants, which survive till the seedling stage, it is completely abolished in knolle keule double mutants that die as large, single-celled multinucleate embryos.

What is the function of the KEULE/KNOLLE complex? Sec1 proteins are large and capable of multiple interactions, and thereby may integrate multiple signals. In yeast and animal cells, Sec1s couple the membrane fusion machinery on target membranes with the Rab GTP cycle on vesicle membranes. In a similar manner, Sec1 proteins constitute a key link between exocytosis and the functional asymmetry and development of neural synapses. By analogy, an intriguing possibility is that KEULE may integrate cell cycle signals and transduce them to the cytokinetic vesicle fusion machinery by virtue of an interaction with the syntaxin KNOLLE. To this effect, a MAP kinase has been found to regulate the activity of a Rab/GDI complex in animal cells, thereby regulating endocytosis. Furthermore, phosphorylation by CDK5 regulates the nSec1-syntaxin1 interaction in neural cells. Thus, the vesicle trafficking machinery could, for instance, be a target of the numerous Ser/Thr kinases, including MAP kinases, present at the phragmoplast.

A Novel Membrane Compartment at the Plane of Division?

The initial rounds of fusion at the cell equator most likely can be considered as "homotypic" in that they occur between like membranes, that is to say, between Golgi-derived vesicles. The new membrane compartment caused by vesicle fusion undergoes a series of visible alterations, including the appearance of a membrane coat and clathrin-coated pits. Although these changes could formally be accompanied by a change in the identity of the cell plate membranes, they are most simply explained as reflecting membrane recycling from the cell plate. In contrast, fusion of the cell plate with the parental wall requires a heterotypic fusion between cell plate membranes and the plasma membrane. After this event, the membranes that delineate the cell plate are contiguous with the plasma membrane, yet of Golgi-derived origin. In fact, during surface expansion and cell elongation, new endomembranes are also added to the plasma membrane.

As of when does the novel membrane compartment formed at the cell equator acquire a plasma membrane identity, and what is its original identity? Phylogenetic analysis places the syntaxin KNOLLE in a novel class of plant-specific syntaxins, as close to plasma membrane syntaxins as it is to endoplasmic reticulum/Golgi syntaxins, and in the same family as AtSYR1, which has been localized to the plasma membrane; KEULE's closest homologs are animal Sec1s required for exocytosis. Green fluorescent protein fusions of the endo-1, 4-β-glucanase KORRIGAN are targeted to both the plasma membrane and cell plate. Evidence that the cell plate membranes in fact differ from the plasma membrane is provided by the observation that the plasma membrane H_+^-ATPase is excluded from the cell plate. These observations support the view that the cell plate is a novel and distinct membrane compartment arising from a modified form of exocytosis.

Cell Plate Maturation

Cell plate maturation requires callose removal as well as cellulose and pectin deposition. Two genes that influence cell wall assembly, CYT1 and KORRIGAN, appear to be required for cell plate consolidation and maturation. Compared with the cell wall stubs observed as of the first zygotic divisions inkeule and knolle mutants, the cell wall stubs observed in cyt1 and korrigan mutants first appear later in development and have only been documented in vacuolated cells. This raises the question as to whether these cell wall stubs arise during cell expansion rather than cell division. It is, nonetheless, clear that the mechanical strength of cell plates is compromised in both mutant backgrounds. CYT1 encodes a Man-1-phosphate guanylyltransferase, required for N-glycosylation. The 5-fold reduction in cellulose content observed in cyt1mutants readily accounts for its extreme cell wall defects, which include a high callose content and a diffuse distribution of unesterified pectins. Noteworthy implications of these findings are that N-linked glycosylation appears to be required for cellulose biosynthesis, and that other polysaccharides such as callose and pectins may compensate for reduced cellulose levels. Reduced cellulose and altered pectin content have also been observed in korrigan mutants. As an endo-1,4-β-glucanase, KORRIGAN is likely to influence the assembly or loosening of cellulose-xyloglucan microfibrils, and is unlikely to be directly implicated in pectin metabolism. Thus, the data point to feedback mechanisms controlling the pectin composition of the cell wall.

Phragmoplast

The phragmoplast is a plant cell specific structure that forms during late cytokinesis. It serves as a scaffold for cell plate assembly and subsequent formation of a new cell wall separating the two daughter cells. The phragmoplast can only be observed in Phragmoplastophyta, a clade that includes the Coleochaetophyceae, Zygnematophyceae, Mesotaeniaceae, and Embryophyta (land plants). Some algae use another type of microtubule array, a phycoplast, during cytokinesis.

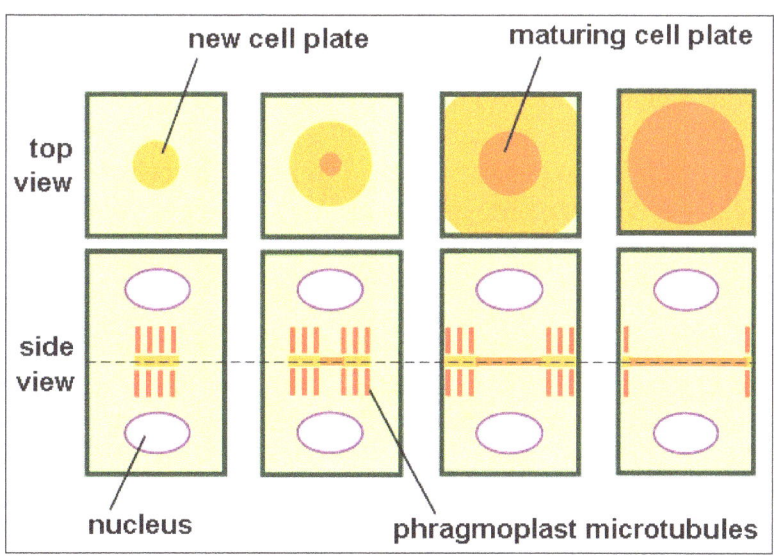

Phragmoplast and cell plate formation in a plant cell during cytokinesis. Left side: Phragmoplast forms and cell plate starts to assemble in the center of the cell. Towards the right: Phragmoplast enlarges in a donut-shape towards the outside of the cell, leaving behind mature cell plate in the center. The cell plate will transform into the new cell wall once cytokinesis is complete.

Structure

The phragmoplast is a complex assembly of microtubules (MTs), microfilaments (MFs), and endo-plasmic reticulum (ER) elements, that assemble in two opposing sets perpendicular to the plane of the future cell plate during anaphase and telophase. It is initially barrel-shaped and forms from the mitotic spindle between the two daughter nuclei while nuclear envelopes reassemble around them. The cell plate initially forms as a disc between the two halves of the phragmoplast structure. While new cell plate material is added to the edges of the growing plate, the phragmoplast mi-crotubules disappear in the center and regenerate at the edges of the growing cell plate. The two structures grow outwards until they reach the outer wall of the dividing cell. If a phragmosome was present in the cell, the phragmoplast and cell plate will grow through the space occupied by the phragmosome. They will reach the parent cell wall exactly at the position formerly occupied by the preprophase band.

The microtubules and actin filaments within the phragmoplast serve to guide vesicles with cell wall material to the growing cell plate. Actin filaments are also possibly involved in guiding the phrag-moplast to the site of the former preprophase band location at the parent cell wall. While the cell plate is growing, segments of smooth endoplasmic reticulum are trapped within it, later forming the plasmodesmata connecting the two daughter cells.

The phragmoplast can be differentiated topographically into two areas, the midline that includes the central plane where some of the plus-ends of both anti-parallel sets of microtubules (MTs) in-terdigitate (as in the midbody matrix), and the distal regions at both sides of the midline.

Role in the Plant Cell Cycle

After anaphase, the phragmoplast emerges from the remnant spindle MTs in between the daugh-ter nuclei. MT plus ends overlap the equator of phragmoplast at the site where the cell plate will form. The formation of the cell plate depends on localized secretory vesicle fusion to deliver mem-brane and cell-wall components. Excess membrane lipid and cell-wall components are recycled by clathrin/dynamin-dependent retrograde membrane traffic. Once the initial cell plate forms at its center, the phragmoplast begins to expand outward to reach the cell edges. Actin filaments also localize to phragmoplast and accumulate greatly at late telophase. Evidence suggests that actin filaments serve phragmoplast expansion more than initial organization, given that disorganization of actin filaments via drug treatments lead to the delay of cell-plate expansion.

Many microtubule-associated proteins (MAPs) have been localized to the phragmoplast, including both constitutively expressed ones (such as MOR1, katanin, CLASP, SPR2, and γ-tubulin complex proteins) and those expressed specifically during M-phase, such as EB1c, TANGLED1 and augmin complex proteins. The functions of these proteins in the phragmoplast are presumably similar to their functions elsewhere in the cell. Most research into phragmoplast MAPs have been focused on the midline because it is, first, where most of the membrane fusion takes place and, second, where the two sets of anti-parallel MTs are held together. The discovery of an important variety of molecules that localize to the phragmoplast midline is shedding light on the complex processes operating in this phragmoplast region.

Two proteins that have critical functions for antiparallel MT bundling at the phragmoplast midline are MAP65-3 and kinesin-5. The kinesin-7 family proteins, HINKEL/AtNACK1 and AtNACK2/TES, recruit a mitogen-activated protein kinase (MAPK) cascade to the midline and induce MAP65 phosphorylation. Phosphorylated MAP65-1 also accumulates at the midline and reduces MT-bundling activities for cell-plate expansion. The essential mechanism of MAPK cascade for phragmoplast expansion is suppressed by cyclin dependent kinase (CDK) activity before telophase.

Certain phragmoplast midline-accumulating MAPs are essential proteins for cytokinesis. The kinesin-12 members, PAKRP1 and PAKRP1L, accumulate at the midline and double loss-of-function mutants have defective cytokinesis during male gametogenesis. PAKRP2 accumulates at midline and also in puncta throughout the phragmoplast, which implies that PAKRP2 participates in Golgi-derived vesicle transport. Moss homologs of PAKRP2, KINID1a, and KINID1b localize to the phragmoplast midline and are essential for phragmoplast organization. RUNKEL, which is a HEAT repeat-containing MAP, also accumulates at the midline and cytokinesis is aberrant in lines with the loss-of-function mutations in this protein. Another midline-localized protein, "two-in-on" (TIO), is a putative kinase and is also required for cytokinesis as shown by defects in a mutant. TIO interacts with PAKRP1, PAKRP1L (kinesin-12), and NACK2/TES (kinesin-7) according to the yeast two hybrid assays. Finally, TPLATE, an adaptin-like protein, accumulates at the cell plate and is essential for cytokinesis.

Stoma

A stoma is a pore, found in the epidermis of leaves, stems, and other organs, that facilitates gas exchange. The pore is bordered by a pair of specialized parenchyma cells known as guard cells that are responsible for regulating the size of the stomatal opening.

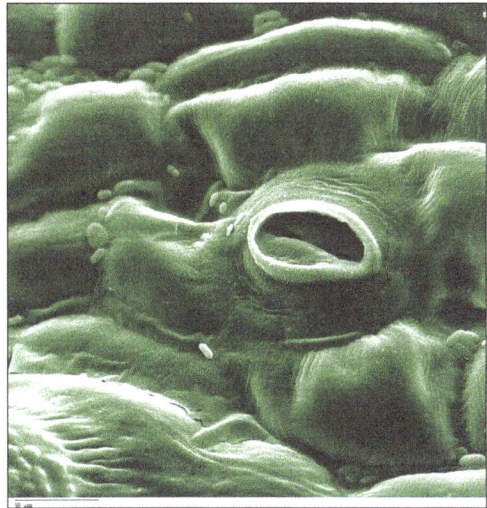

Stoma in a tomato leaf shown via colorized scanning electron microscope image.

The term is usually used collectively to refer to the entire stomatal complex, consisting of the paired guard cells and the pore itself, which is referred to as the stomatal aperture. Air enters the plant through these openings by gaseous diffusion and contains carbon dioxide which is used in photosynthesis and oxygen which is used in respiration. Oxygen produced as a by-product of photosynthesis diffuses out to the atmosphere through these same openings. Also, water vapor diffuses through the stomata into the atmosphere in a process called transpiration.

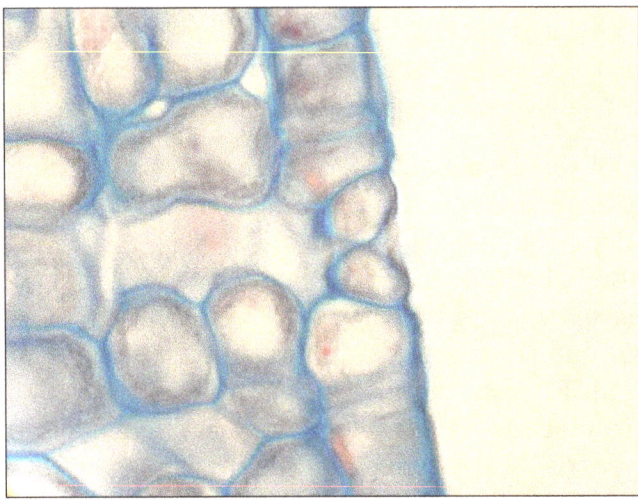

A stoma in cross section.

Stomata are present in the sporophyte generation of all land plant groups except liverworts. In vascular plants the number, size and distribution of stomata varies widely. Dicotyledons usually have more stomata on the lower surface of the leaves than the upper surface. Monocotyledons such as onion, oat and maize may have about the same number of stomata on both leaf surfaces. In plants with floating leaves, stomata may be found only on the upper epidermis and submerged leaves may lack stomata entirely. Most tree species have stomata only on the lower leaf surface. Leaves with stomata on both the upper and lower leaf are called amphistomatous leaves; leaves with stomata only on the lower surface are hypostomatous, and leaves with stomata only on the upper surface are epistomatous or hyperstomatous. Size varies across species, with end-to-end lengths ranging from 10 to 80 µm and width ranging from a few to 50 µm.

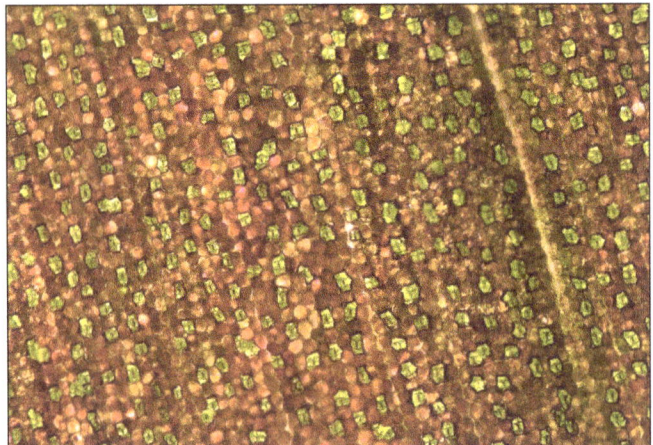

The underside of a leaf. In this species (Tradescantia zebrina) the guard cells of the stomata are green because they contain chlorophyll while the epidermal cells are chlorophyll-free and contain red pigments.

Function

CO₂ Gain and Water Loss

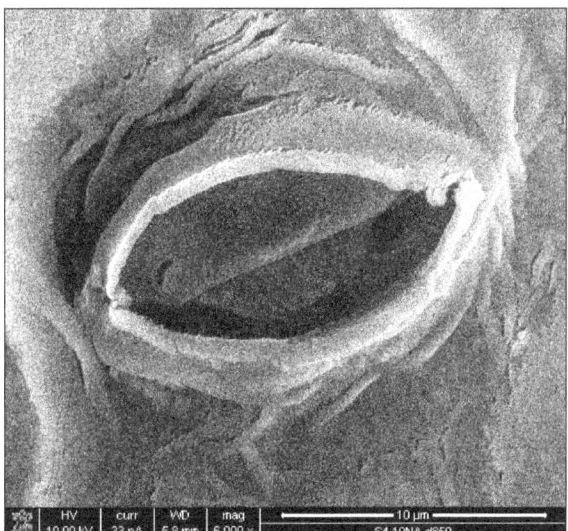

Electron micrograph of a stoma from a Brassica chinensis (Bok Choy) leaf.

Carbon dioxide, a key reactant in photosynthesis, is present in the atmosphere at a concentration of about 400 ppm. Most plants require the stomata to be open during daytime. The air spaces in the leaf are saturated with water vapour, which exits the leaf through the stomata in a process known as transpiration. Therefore, plants cannot gain carbon dioxide without simultaneously losing water vapour.

Alternative Approaches

Ordinarily, carbon dioxide is fixed to ribulose-1,5-bisphosphate (RuBP) by the enzyme RuBisCO in mesophyll cells exposed directly to the air spaces inside the leaf. This exacerbates the transpiration problem for two reasons: first, RuBisCo has a relatively low affinity for carbon dioxide, and second, it fixes oxygen to RuBP, wasting energy and carbon in a process called photorespiration. For both of these reasons, RuBisCo needs high carbon dioxide concentrations, which means wide stomatal apertures and, as a consequence, high water loss.

Narrower stomatal apertures can be used in conjunction with an intermediary molecule with a high carbon dioxide affinity, PEPcase (Phosphoenolpyruvate carboxylase). Retrieving the products of carbon fixation from PEPCase is an energy-intensive process, however. As a result, the PEPCase alternative is preferable only where water is limiting but light is plentiful, or where high temperatures increase the solubility of oxygen relative to that of carbon dioxide, magnifying RuBisCo's oxygenation problem.

CAM Plants

A group of mostly desert plants called "CAM" plants (Crassulacean acid metabolism, after the family Crassulaceae, which includes the species in which the CAM process was first discovered) open their stomata at night (when water evaporates more slowly from leaves for a given degree of stomatal opening), use PEPcarboxylase to fix carbon dioxide and store the products in large vacuoles. The following day, they close their stomata and release the carbon dioxide fixed the previous night into the presence of RuBisCO. This saturates RuBisCO with carbon dioxide, allowing minimal photorespiration. This approach, however, is severely limited by the capacity to store fixed carbon in the vacuoles, so it is preferable only when water is severely limited.

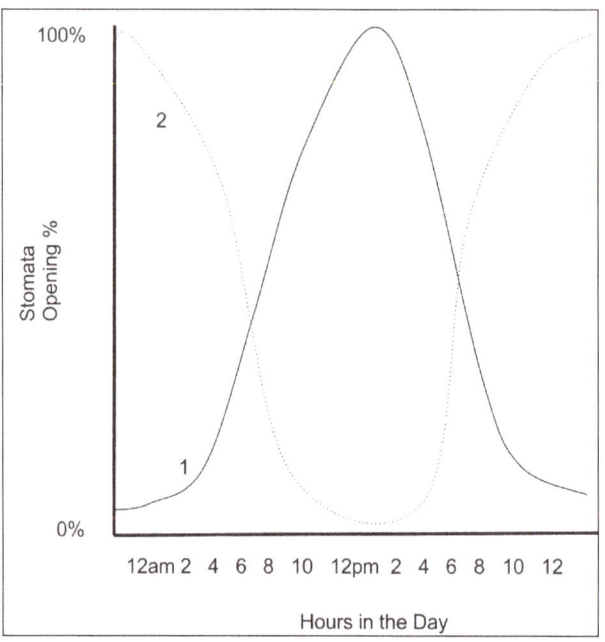

C3 and C4 plants(1) stomata stay open all day and close at night. CAM plants(2) stomata open during the morning and close slightly at noon and then open again in the evening.

Opening and Closing

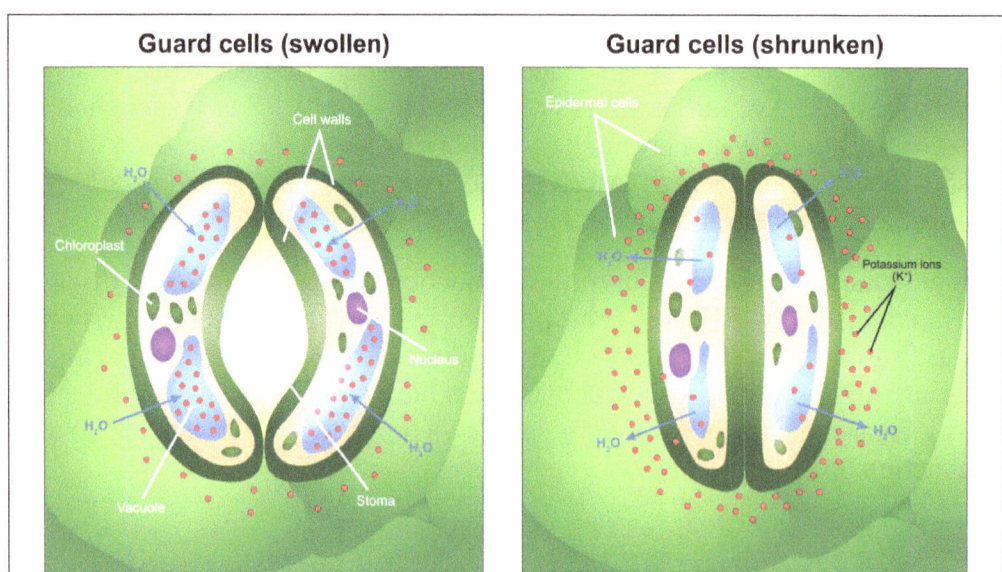

Opening and closing of stoma.

However, most plants do not have the aforementioned facility and must therefore open and close their stomata during the daytime, in response to changing conditions, such as light intensity, humidity, and carbon dioxide concentration. It is not entirely certain how these responses work. However, the basic mechanism involves regulation of osmotic pressure.

When conditions are conducive to stomatal opening (e.g., high light intensity and high humidity), a proton pump drives protons (H^+) from the guard cells. This means that the cells' electrical potential becomes increasingly negative. The negative potential opens potassium voltage-gated channels and so an uptake of potassium ions (K^+) occurs. To maintain this internal negative voltage so that entry of potassium ions does not stop, negative ions balance the influx of potassium. In some cases, chloride ions enter, while in other plants the organic ion malate is produced in guard cells. This increase in solute concentration lowers the water potential inside the cell, which results in the diffusion of water into the cell through osmosis. This increases the cell's volume and turgor pressure. Then, because of rings of cellulose microfibrils that prevent the width of the guard cells from swelling, and thus only allow the extra turgor pressure to elongate the guard cells, whose ends are held firmly in place by surrounding epidermal cells, the two guard cells lengthen by bowing apart from one another, creating an open pore through which gas can move.

When the roots begin to sense a water shortage in the soil, abscisic acid (ABA) is released. ABA binds to receptor proteins in the guard cells' plasma membrane and cytosol, which first raises the pH of the cytosol of the cells and cause the concentration of free Ca^{2+} to increase in the cytosol due to influx from outside the cell and release of Ca^{2+} from internal stores such as the endoplasmic reticulum and vacuoles. This causes the chloride (Cl^-) and organic ions to exit the cells. Second, this stops the uptake of any further K^+ into the cells and, subsequently, the loss of K^+. The loss of these solutes causes an increase in water potential, which results in the diffusion of water back out of the cell by osmosis. This makes the cell plasmolysed, which results in the closing of the stomatal pores.

Guard cells have more chloroplasts than the other epidermal cells from which guard cells are derived. Their function is controversial.

Inferring Stomatal Behavior from Gas Exchange

The degree of stomatal resistance can be determined by measuring leaf gas exchange of a leaf. The transpiration rate is dependent on the diffusion resistance provided by the stomatal pores, and also on the humidity gradient between the leaf's internal air spaces and the outside air. Stomatal resistance (or its inverse, stomatal conductance) can therefore be calculated from the transpiration rate and humidity gradient. This allows scientists to investigate how stomata respond to changes in environmental conditions, such as light intensity and concentrations of gases such as water vapor, carbon dioxide, and ozone. Evaporation (E) can be calculated as:

$$E = (e_i - e_a)/Pr$$

where e_i and e_a are the partial pressures of water in the leaf and in the ambient air, respectively, P is atmospheric pressure, and r is stomatal resistance. The inverse of r is conductance to water vapor (g), so the equation can be rearranged to:

$$E = (e_i - e_a)g/P$$

and solved for g:

$$g = EP / (e_i - e_a)$$

Photosynthetic CO_2 assimilation (A) can be calculated from:

$$A = (C_a - C_i)g / 1.6P$$

where C_a and C_i are the atmospheric and sub-stomatal partial pressures of CO_2, respectively. The rate of evaporation from a leaf can be determined using a photosynthesis system. These scientific instruments measure the amount of water vapour leaving the leaf and the vapor pressure of the ambient air. Photosynthetic systems may calculate water use efficiency (A/E), g, intrinsic water use efficiency (A/g), and C_i. These scientific instruments are commonly used by plant physiologists to measure CO_2 uptake and thus measure photosynthetic rate.

Development

There are three major epidermal cell types which all ultimately derive from the outermost (L1) tissue layer of the shoot apical meristem, called protodermal cells: trichomes, pavement cells and guard cells, all of which are arranged in a non-random fashion.

An asymmetrical cell division occurs in protodermal cells resulting in one large cell that is fated to become a pavement cell and a smaller cell called a meristemoid that will eventually differentiate into the guard cells that surround a stoma. This meristemoid then divides asymmetrically one to three times before differentiating into a guard mother cell. The guard mother cell then makes one symmetrical division, which forms a pair of guard cells. Cell division is inhibited in some cells so there is always at least one cell between stomata.

Stomatal patterning is controlled by the interaction of many signal transduction components such as EPF (Epidermal Patterning Factor), ERL (ERecta Like) and YODA (a putative MAP kinase kinase kinase). Mutations in any one of the genes which encode these factors may alter the development of stomata in the epidermis. For example, a mutation in one gene causes more stomata that are clustered together, hence is called Too Many Mouths (TMM). Whereas, disruption of the SPCH (SPeec-CHless) gene prevents stomatal development all together. Activation of stomatal production can occur by the activation of EPF1, which activates TMM/ERL, which together activate YODA. YODA inhibits SPCH, causing SPCH activity to decrease, allowing for asymmetrical cell division that initiates stomata formation. Stomatal development is also coordinated by the cellular peptide signal called stomagen, which signals the inhibition of the SPCH, resulting in increased number of stomata.

Environmental and hormonal factors can affect stomatal development. Light increases stomatal development in plants; while, plants grown in the dark have a lower amount of stomata. Auxin represses stomatal development by affecting their development at the receptor level like the ERL and TMM receptors. However, a low concentration of auxin allows for equal division of a guard mother cell and increases the chance of producing guard cells.

Types

Different classifications of stoma types exist. One that is widely used is based on the types that

Julien Joseph Vesque introduced in 1889, was further developed by Metcalfe and Chalk, and later complemented by other authors. It is based on the size, shape and arrangement of the subsidiary cells that surround the two guard cells. They distinguish for dicots:

- Actinocytic (meaning star-celled) stomata have guard cells that are surrounded by at least five radiating cells forming a star-like circle. This is a rare type that can for instance be found in the family Ebenaceae.

- Anisocytic (meaning unequal celled) stomata have guard cells between two larger subsidiary cells and one distinctly smaller one. This type of stomata can be found in more than thirty dicot families, including Brassicaceae, Solanaceae, and Crassulaceae. It is sometimes called cruciferous type.

- Anomocytic (meaning irregular celled) stomata have guard cells that are surrounded by cells that have the same size, shape and arrangement as the rest of the epidermis cells. This type of stomata can be found in more than hundred dicot families such as Apocynaceae, Boraginaceae, Chenopodiaceae, and Cucurbitaceae. It is sometimes called ranunculaceous type.

- Diacytic (meaning cross-celled) stomata have guard cells surrounded by two subsidiary cells, that each encircle one end of the opening and contact each other opposite to the middle of the opening. This type of stomata can be found in more than ten dicot families such as Caryophyllaceae and Acanthaceae. It is sometimes called caryophyllaceous type.

- Hemiparacytic stomata are bordered by just one subsidiary cell that differs from the surrounding epidermis cells, its length parallel to the stoma opening. This type occurs for instance in the Molluginaceae and Aizoaceae.

- Paracytic (meaning parallel celled) stomata have one or more subsidiary cells parallel to the opening between the guard cells. These subsidiary cells may reach beyond the guard cells or not. This type of stomata can be found in more than hundred dicot families such as Rubiaceae, Convolvulaceae and Fabaceae. It is sometimes called rubiaceous type.

In monocots, several different types of stomata occur such as:

- Gramineous (meaning grass-like) stomata have two guard cells surrounded by two lens-shaped subsidiary cells. The guard cells are narrower in the middle and bulbous on each end. This middle section is strongly thickened. The axis of the subsidiary cells are parallel stoma opening. This type can be found in monocot families including Poaceae and Cyperaceae.

- Hexacytic (meaning six-celled) stomata have six subsidiary cells around both guard cells, one at either end of the opening of the stoma, one adjoining each guard cell, and one between that last subsidiary cell and the standard epidermis cells. This type can be found in some monocot families.

- Tetracytic (meaning four-celled) stomata have four subsidiary cells, one on either end of the opening, and one next to each guard cell. This type occurs in many monocot families, but also can be found in some dicots, such as Tilia and several Asclepiadaceae.

In ferns, four different types are distinguished:

- Hypocytic stomata have two guard cells in one layer with only ordinary epidermis cells, but with two subsidiary cells on the outer surface of the epidermis, arranged parallel to the guard cells, with a pore between them, overlying the stoma opening.

- Pericytic stomata have two guard cells that are entirely encircled by one continuous subsidiary cell (like a donut).

- Desmocytic stomata have two guard cells that are entirely encircled by one subsidiary cell that has not merged its ends (like a sausage).

- Polocytic stomata have two guard cells that are largely encircled by one subsidiary cell, but also contact ordinary epidermis cells (like a U or horseshoe).

Stomatal Crypts

Stomatal crypts are sunken areas of the leaf epidermis which form a chamber-like structure that contains one or more stomata and sometimes trichomes or accumulations of wax. Stomatal crypts can be an adaption to drought and dry climate conditions when the stomatal crypts are very pronounced. However, dry climates are not the only places where they can be found. The following plants are examples of species with stomatal crypts or antechambers: Nerium oleander, conifers, and Drimys winteri which is a species of plant found in the cloud forest.

Stomata as Pathogenic Pathways

Stomata are obvious holes in the leaf by which, as was presumed for a while, pathogens can enter unchallenged. However, it has been recently shown that stomata do in fact sense the presence of some, if not all, pathogens. However, with the virulent bacteria applied to Arabidopsis plant leaves in the experiment, the bacteria released the chemical coronatine, which forced the stomata open again within a few hours.

Stomata and Climate Change

Response of Stomata to Environmental Factors

Photosynthesis, plant water transport (xylem) and gas exchange are regulated by stomatal function which is important in the functioning of plants. Stomatal density and aperture (length of stomata) varies under a number of environmental factors such as atmospheric CO_2 concentration, light intensity, air temperature and photoperiod (daytime duration).

Decreasing stomatal density is one way plants have responded to the increase in concentration of atmospheric CO_2 ($[CO_2]_{atm}$). Although changes in $[CO_2]_{atm}$ response is the least understood mechanistically, this stomatal response has begun to plateau where it is soon expected to impact transpiration and photosynthesis processes in plants.

Future Adaptations during Climate Change

It is expected for $[CO_2]_{atm}$ to reach 500–1000 ppm by 2100. 96% of the past 400 000 years experienced below 280 ppm CO_2 levels. From this figure, it is highly probable that genotypes of today's plants diverged from their pre-industrial relative.

The gene HIC (high carbon dioxide) encodes a negative regulator for the development of stomata in plants. Research into the HIC gene using Arabidopsis thaliana found no increase of stomatal development in the dominant allele, but in the 'wild type' recessive allele showed a large increase, both in response to rising CO_2 levels in the atmosphere. These studies imply the plants response to changing CO_2 levels is largely controlled by genetics.

Agricultural Implications

The CO_2 fertiliser effect has been greatly overestimated during Free-Air Carbon dioxide Enrichment (FACE) experiments where results show increased CO_2 levels in the atmosphere enhances photosynthesis, reduce transpiration, and increase water use efficiency (WUE). Increased biomass is one of the effects with simulations from experiments predicting a 5–20% increase in crop yields at 550 ppm of CO_2. Rates of leaf photosynthesis were shown to increase by 30–50% in C3 plants, and 10–25% in C4 under doubled CO_2 levels. The existence of a feedback mechanism results a phenotypic plasticity in response to $[CO_2]_{atm}$ that may have been an adaptive trait in the evolution of plant respiration and function.

Predicting how stomata perform during adaptation is useful for understanding the productivity of plant systems for both natural and agricultural systems. Plant breeders and farmers are beginning to work together using evolutionary and participatory plant breeding to find the best suited species such as heat and drought resistant crop varieties that could naturally evolve to the change in the face of food security challenges.

Endoplasmic Reticulum

The endoplasmic reticulum (ER) is a large organelle made of membranous sheets and tubules that begin near the nucleus and extend across the cell. The endoplasmic reticulum creates, packages, and secretes many of the products created by a cell. Ribosomes, which create proteins, line a portion of the endoplasmic reticulum.

The entire structure can account for a large proportion of the endomembrane system of the cell. For instance, in cells such as liver hepatocytes that are specialized for protein secretion and detoxification, the ER can account for more than 50% of the total lipid bilayer of the cell. Similarly, the ER membrane system is particularly prominent in pancreatic beta cells that secrete insulin, or within activated B-lymphocytes that produce antibodies.

As seen in the image, the membranes of the endoplasmic reticulum are contiguous with the outer nuclear membrane, even though their compositions can be different. The ER contains special membrane-embedded proteins that stabilize its structure and curvature. This organelle acts as an important regulator of cell function because it interacts closely with a number of other organelles. Products of the endoplasmic reticulum often travel to the Golgi body for packaging and additional processing before being secreted.

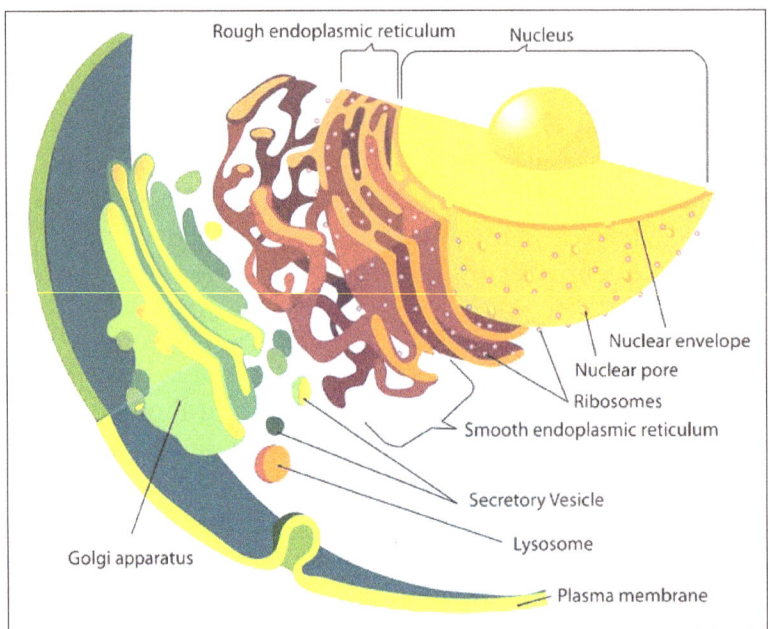

Endomembrane system diagram.

This is a microscopic image of a section from mammalian lung tissue. The bottom right corner of the image shows the nucleus and the rest of the picture illustrates the extensive nature of the ER. Small dark circles are mitochondria that exist in physical proximity with the membranes of the ER.

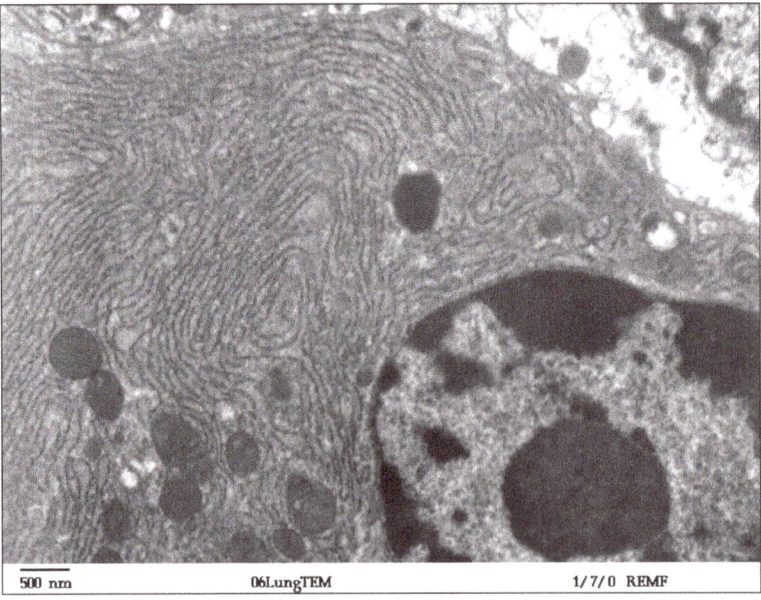

Mammalian lung tissue.

Endoplasmic Reticulum Function

The ER plays a number of roles within the cell, from protein synthesis and lipid metabolism to detoxification of the cell. Cisternae, each of the small folds of the endoplasmic reticulum, are commonly associated with lipid metabolism. This creates the plasma membrane of the cell, as well as additional endoplasmic reticulum and organelles. They also appear to be important in maintaining the Ca^{2+} balance within the cell and in the interaction of the ER with mitochondria. This interaction also influences the aerobic status of the cell.

ER sheets appear to be crucial in the response of the organelle to stress, especially since cells alter their tubules-to-sheets ratio when the number of unfolded proteins increases. Occasionally, apoptosis is induced by the ER in response to an excess of unfolded protein within the cell. When ribosomes detach from ER sheets, these structures can disperse and form tubular cisternae.

Although ER sheets and tubules appear to have distinct functions, there isn't a perfect delineation of roles. For instance, in mammals tubules and sheets can interconvert, making the cells adaptable to various conditions. The relationship between structure and function in the ER has not been completely elucidated.

Protein Synthesis and Folding

Protein synthesis occurs in the rough endoplasmic reticulum. Although translation for all proteins begins in the cytoplasm, some are moved into the ER in order to be folded and sorted for different destinations. Proteins that are translocated into the ER during translation are often destined for secretion. Initially, these proteins are folded within the ER and then moved into the Golgi apparatus where they can be dispatched towards other organelles.

For instance, the hydrolytic enzymes in the lysosome are generated in this manner. Alternately, these proteins could be secreted from the cell. This is the origin of the enzymes of the digestive tract. The third potential role for proteins translated in the ER is to remain within the endomembrane system itself. This is particularly true for chaperone proteins that assist in the folding of other proteins. The genes encoding these proteins are unregulated when the cell is under stress from unfolded proteins.

Lipid Synthesis

The smooth endoplasmic reticulum plays an important role in cholesterol and phospholipid biosynthesis. Therefore, ER is important not only for the generation and maintenance of the plasma membrane but of the extensive endomembrane system of the ER itself.

The SER is enriched in enzymes involved in sterol and steroid biosynthetic pathways and is also necessary for the synthesis of steroid hormones. Therefore the SER is extremely prominent in the cells of the adrenal gland that secrete five different groups of steroid hormones that influence the metabolism of the entire body. The synthesis of these hormones also involves enzymes within the mitochondria, further underscoring the relationship between these two organelles.

Calcium Store

The SER is an important site for the storage and release of calcium in the cell. A modified form of the SER called sarcoplasmic reticulum forms an extensive network in contractile cells such as muscle fibers. Calcium ions are also involved in the regulation of metabolism in the cell and can change cytoskeletal dynamics.

The extensive nature of the ER network allows it to interact with the plasma membrane and use

Ca²⁺ for signal transduction and modulation of nuclear activity. In association with mitochondria, the ER can also use its calcium stores to induce apoptosis in response to stress.

Structure of the Endoplasmic Reticulum

The endoplasmic reticulum membrane system can be morphologically divided into two structures–cisternae and sheets. Cisternae are tubular in structure and form a three-dimensional polygonal network. They are about 50 nm in diameter in mammals and 30 nm in diameter in yeast. ER sheets, on the other hand, are membrane-enclosed, two-dimensional flattened sacs that extend across the cytoplasm. They are frequently associated with ribosomes and special proteins called translocons that are necessary for protein translation within the RER.

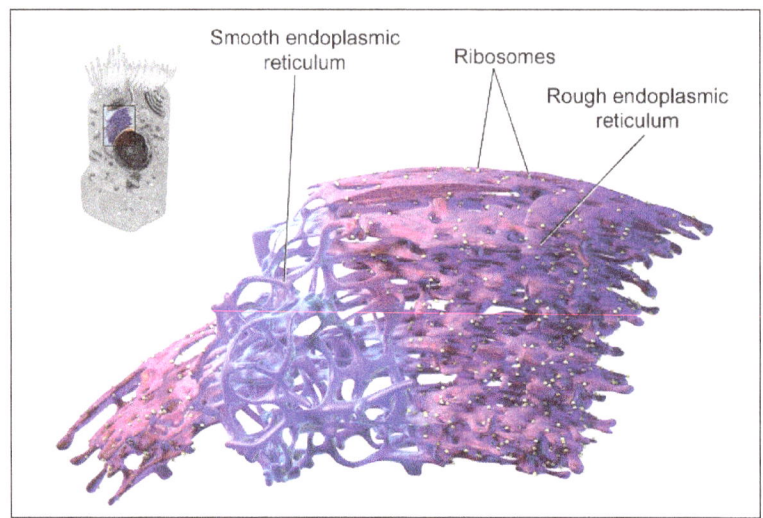

Endoplasmic Reticulum Structure.

The high-curvature of ER tubules is stabilized by the presence of proteins called reticulons and DP1/Yop1p. Reticulons are membrane-associated proteins encoded by four genes in mammals (RTN1-4). These proteins localize to ER tubules and the curved edges of ER sheets. DP1/Yop1p are a class of integral membrane proteins involved in stabilizing the structure of ER cisternae.

Both reticulons and DP1/Yop1 proteins form oligomers and interact with the cytoskeleton. Oligomerization seems to be one of the mechanisms used by these proteins to shape the lipid bilayer into a tubule. Additionally, they also appear to use a wedge-like structural motif that causes the membrane to curve. These two classes of proteins are redundant, since the overexpression of one protein can compensate for the lack of the other protein.

The construction of the ER is intimately involved with the presence of cytoskeletal elements, especially microtubules. ER membranes, especially cisternae, move and branch along microtubules. When microtubule structure is temporarily disrupted, the ER network collapses and reforms only after the microtubule cytoskeleton is reestablished. In addition, changes to the pattern of microtubule polymerization are reflected in changes to ER morphology.

Endoplasmic Reticulum Location

The endoplasmic reticulum processes most of the instructions from the nucleus. As such, the endoplasmic reticulum surrounds the nucleus and radiates outward. In cells that secrete many products for the rest of the body, the endoplasmic reticulum can account for more than 50% of the cell.

In general, the nucleus expresses mRNA (messenger RNA), which tells the cell how to build proteins. The rough endoplasmic reticulum has many ribosomes, which are the primary location of protein production. This portion of the organelle creates proteins and begins to fold them into the proper formation. The smooth endoplasmic reticulum is the primary location for lipid synthesis. As such, it does not contain any ribosomes. Rather, it conducts a series of reactions which create the phospholipid molecules necessary to create various membranes and organelles.

The rough version of the endoplasmic reticulum is often closer to the nucleus, whereas the smooth endoplasmic reticulum is further from the nucleus. However, both versions are connected to each other and the nucleus through a series of small tubules.

There are two types of endoplasmic reticulum: rough endoplasmic reticulum (rough ER) and smooth endoplasmic reticulum (smooth ER). Both types are present in plant and animal cells. The two types of ER often appear as if separate, but they are sub-compartments of the same organelle. Cells specialising in the production of proteins will tend to have a larger amount of rough ER whilst cells producing lipids (fats) and steroid hormones will have a greater amount of smooth ER.

Part of the ER is contiguous with the nuclear envelope. The Golgi apparatus is also closely associated with the ER and recent observations suggest that parts of the two organelles, i.e. the ER and the Golgi complex, are so close that some chemical products probably pass directly between them instead of being packaged into vesicles (droplets enclosed within a membrane) and transported to them through the cytoplasm.

Rough Endoplasmic Reticulum

This is an extensive organelle composed of greatly convoluted but flattish sealed sacs, which are contiguous with the nuclear membrane. It is called 'rough' endoplasmic reticulum because it is studded on its outer surface (the surface in contact with the cytosol) with ribosomes. These are called membrane bound ribosomes and are firmly attached to the outer cytosolic side of the ER About 13 million ribosomes are present on the RER in the average liver cell. Rough ER is found throughout the cell but the density is higher near the nucleus and the Golgi apparatus.

Ribosomes on the rough endoplasmic reticulum are called 'membrane bound' and are responsible for the assembly of many proteins. This process is called translation. Certain cells of the pancreas and digestive tract produce a high volume of protein as enzymes. Many of the proteins are produced in quantity in the cells of the pancreas and the digestive tract and function as digestive enzymes.

The rough ER working with membrane bound ribosomes takes polypeptides and amino acids from the cytosol and continues protein assembly including, at an early stage, recognising a 'destination label' attached to each of them. Proteins are produced for the plasma membrane, Golgi apparatus, secretory vesicles, plant vacuoles, lysosomes, endosomes and the endoplasmic reticulum itself. Some of the proteins are delivered into the lumen or space inside the ER whilst others are processed within the ER membrane itself. In the lumen some proteins have sugar groups added to them to form glycoproteins. Some have metal groups added to them. It is in the rough ER for example that four polypeptide chains are brought together to form haemoglobin.

Protein Folding Unit

It is in the lumen of the rough ER that proteins are folded to produce the highly important bio-chemical architecture which will provide 'lock and key' and other recognition and linking sites.

Protein Quality Control Section

It is also in the lumen that an amazing process of quality control checking is carried out. Proteins are subjected to a quality control check and any that are found to be incorrectly formed or incor-rectly folded are rejected. These rejects are stored in the lumen or sent for recycling for eventual breakdown to amino acids. A type of emphysema (a lung problem) is caused by the ER quality control section continually rejecting an incorrectly folded protein. The protein is wrongly folded as a result of receiving an altered genetic message. The required protein is never exported from the lumen of rough ER. Research into protein structure failures relating to HIV are also focusing on reactions in the ER.

Rigorous Quality Control Plays a Part in Cystic Fibrosis

A form of cystic fibrosis is caused by a missing single amino acid, phenylanaline, in a particular position in the protein construction. The protein might work well without the amino acid but the very exacting service provided by the quality control section spots the error and rejects the protein retaining it in the lumen of the rough ER. In this case the customer (the person with cystic fibrosis) loses out completely due to high standards when a slightly poorer product would have been better than no product at all.

From Rough ER to Golgi

In most cases proteins are transferred to the Golgi apparatus for 'finishing'. They are conveyed in vesicles or possibly directly between the ER and Golgi surfaces. After 'finishing' they are delivered to specific locations.

Smooth Endoplasmic Reticulum

Smooth ER is more tubular than rough ER and forms an interconnecting network sub-compart-ment of ER. It is found fairly evenly distributed throughout the cytoplasm.

It is not studded with ribosomes hence 'smooth' ER.

Smooth ER is devoted almost exclusively to the manufacture of lipids and in some cases to the me-tabolism of them and associated products. In liver cells for example smooth ER enables glycogen that is stored as granules on the external surface of smooth ER to be broken down to glucose. Smooth ER is also involved in the production of steroid hormones in the adrenal cortex and endocrine glands.

Smooth ER also plays a large part in detoxifying a number of organic chemicals converting them to safer water-soluble products.

Large amounts of smooth ER are found in liver cells where one of its main functions is to detoxify products of natural metabolism and to endeavour to detoxify overloads of ethanol derived from excess alcoholic drinking and also barbiturates from drug overdose. To assist with this, smooth ER can double its surface area within a few days, returning to its normal size when the assault has subsided.

The contraction of muscle cells is triggered by the orderly release of calcium ions. These ions are released from the smooth endoplasmic reticulum.

Plant Cell Vacuoles

The vacuoles of plant cells are multifunctional organelles that are central to cellular strategies of plant development. They share some of their basic properties with the vacuoles of algae and yeast and the lysosomes of animal cells. They are lytic compartments, function as reservoirs for ions and metabolites, including pigments, and are crucial to processes of detoxification and general cell homeostasis. They are involved in cellular responses to environmental and biotic factors that provoke stress. In the vegetative organs of the plant, they act in combination with the cell wall to generate turgor, the driving force for hydraulic stiffness and growth. In seeds and specialized storage tissues, they serve as sites for storing reserve proteins and soluble carbohydrates. In this way, vacuoles serve physical and metabolic functions that are essential to plant life.

Plant cell vacuoles were discovered with the early microscope and, as indicated in the etymology of the word, originally defined as a cell space empty of cytoplasmic matter. Technical progress has variously altered the operating definition of the plant vacuole over time. Today, definitions continue to be colored by the tools and concepts brought to bear in any given study. Indeed, the combination of microscopy, biochemistry, genetics, and molecular biology is fundamental to research into the plant vacuole.

Vacuoles are provisionally defined as the intracellular compartments that arise as a terminal product of the secretory pathway in plant cells. They are ontogenetically and functionally linked with other components of the vacuolar apparatus (i.e., vacuoles and those membranous bodies that are either committed to becoming vacuolar or have immediately completed a vacuolar function). Experimental evidence suggests that material within the vacuolar system in plants derives confluently from both an intracellular biosynthetic pathway and a coordinated endocytotic pathway. The biogenetic pathways include (1) sorting of proteins destined for the vacuole away from those to be delivered to the cell surface after transit through the early stages of the secretory pathway; (2) endocytosis of materials from the plasma membrane; (3) autophagy pathways for vacuole formation; and (4) direct cytoplasm-to-vacuole delivery. Ultimately, sorting and targeting mechanisms ensure that specific proteins are faithfully assigned to conduct the vacuolar functions.

The Diversity of Vacuoles

Plant cell vacuoles are widely diverse in form, size, content, and functional dynamics, and a single cell may contain more than one kind of vacuole. Although major morphological differences were recorded by the very first microscopists, it has been commonly assumed that all vacuoles have the same origin and belong to a common group. However, with improvements in cell fractionation and biochemical analyses as well as in the use of new molecular probes, it has become possible to characterize specialized vacuolar compartments in the cells from a variety of tissues.

In most cells from the vegetative tissues of the plant body, the central vacuole occupies much of the volume and is essential for much of the physiology of the organism. Among the many functions of this organelle are turgor maintenance, protoplasmic homeostasis, storage of metabolic products, sequestration of xenobiotics, and digestion of cytoplasmic constituents. In regard to the latter function, vacuoles are acidic and contain hydrolytic enzymes analogous to the lysosomal enzymes of animal cells. The membrane, or tonoplast, of such vacuoles contains the vegetative-specific aquaporin γ-TIP. In some cell types, defense or signal compounds are stored in the vacuole, particularly within specialized cells located in strategically favorable tissues such as the leaf epidermis. As early as last century, it was observed that many pigments (e.g., anthocyanins) are localized in the vacuoles of epidermal cells from flowers, leaves, and stems. Recent findings suggest that the membranes of such specialized vacuoles contain specific ATP binding cassette (ABC) transporters.

In contrast, reserve tissues of seeds and fruit contain vacuoles specialized in the storage of proteins. The membrane of the protein storage vacuoles (PSVs) contains the seed-specific aquaporin α-TIP. Storage proteins are also synthesized and accumulated in specialized vegetative cells in response to wounding and to developmental switches. Distinctively, the membrane of the vegetative storage vacuoles contains the aquaporin ∂-TIP. In the endosperm of cereal grains, proteins accumulate in endoplasmic reticulum (ER)–derived organelles of vacuole-like size.

A few recent studies show that distinct vacuoles may simultaneously function in the same cell. Two separate vacuolar compartments, defined by α-TIP and γ-TIP, occur together in the root tip cells of barley and pea seedlings, mature tobacco plants, as well as in the plumule cells of pea seedlings. Barley lectin in root tip cells is found within α-TIP–positive vacuoles but not in γ-TIP–positive vacuoles, whereas the barley acid cysteine protease, aleurain, is specifically contained within γ-TIP–positive vacuoles but is absent from α-TIP–positive vacuoles. Thus, α-TIP defines a storage vacuole in which proteins are protected against degradative enzymes, whereas γ-TIP defines a separate, acidic, lytic vacuole. As cells develop large vacuoles, these two compartments appear to merge because the marker membrane antigens, α-TIP and γ-TIP, colocalize to the same membrane, at least in certain regions of the vacuolar compartments.

Two distinct vacuole types are similarly found in living protoplasts of barley aleurone. In addition to PSVs, aleurone cells contain a second type of lytic organelle, designated as secondary vacuoles by the authors of this study. Although PSVs and secondary vacuoles are lytic organelles with acidic contents, it was suggested that the secondary vacuoles, which have many features typical of plant vacuoles, function as lysosomes and could be involved in the programmed death of aleurone cells.

Another example of the versatility of vacuoles comes from investigations of the motor cells of the pulvini from Mimosa pudica . The vacuole that occurs in the immature (nonreactive) motor cell is located near the nucleus, contains large amounts of tannins, and is believed to act as a Ca^{2+} store. The "aqueous" vacuole that is additionally found in mature motor cells does not contain tannins, is much larger than the tannin-containing vacuole, and occupies a central position in mature cells. The changes in cell volume that are responsible for pulvini-mediated leaf movement result from massive water fluxes mainly across the membrane of the larger, aqueous vacuole, on which the γ-TIP aquaporin and the vacuolar-type H $^+$-translocating ATPase (V-ATPase) are detected almost exclusively. Both vacuoles change shape to effect cell shrinkage. The tannin-containing vacuole forms interconnected tubules, whereas the aqueous vacuole develops membrane wrinkles. In any case, the tannin vacuole and the aqueous vacuole do not merge but rather coexist within the mature motor cell.

Additionally, because vacuoles are highly dynamic organelles, often capable of transforming in terms of both form and function, several "generations" of vacuole may be found within a given cell. In the cells of developing pea cotyledons, for instance, two categories of vacuole are reported: a declining, vegetative γ-TIP–associated vacuole; and a newly formed, α-TIP–associated storage vacuole. Moreover, in suspension-cultured cells subjected to sucrose starvation, protein degradation is supported by numerous active autophagic vacuoles that are present together with the large, more mature central vacuole. After completion of autophagic digestion, the small vacuoles are subsequently incorporated into the central vacuole. However, when intracellular digestion is inhibited, autophagic vacuoles containing undigested substrates remain in the cytoplasm as residual bodies, apart from the large central vacuole.

The diversity of function and form outlined in the above examples illustrates that the cytological definition of vacuoles is likely to cover several biochemically and physiologically distinct entities. Vacuoles, as dynamic organelles, can thus be viewed in the right perspective only if their dynamic nature itself is understood. In several instances, entities that may be variously defined according to different morphological, biochemical, and physical criteria may not necessarily correspond to distinct physiological units.

Biogenesis of Vegetative Vacuoles

Until recently, our knowledge of the biosynthesis and maintenance of vacuoles was based largely on morphological observations. Technological breakthroughs over the past few years have advanced our understanding of vacuolar biogenesis to a more detailed molecular level. Resident vacuolar proteins as well as proteins destined for degradation are delivered to the vacuole via the secretory pathway, which includes the biosynthetic, autophagic, and endocytotic transport routes that are presented in figure. The basic mechanisms that organize these routes in eukaryotes are highly conserved across phyla.

Early Secretory Pathway

In plant cells, as in animal cells and yeast, anterograde transport of newly synthesized soluble as well as membrane proteins through the vacuolar pathway begins at the ER. Most soluble proteins destined for the vacuole are synthesized as precursors with a transient N-terminal signal peptide by membrane-bound polysomes. The nascent precursor form is efficiently targeted to the ER lumen. After their cotranslational translocation across the ER membrane, the secretory proteins are folded and subjected to early post-translational modifications. ER-resident proteins, such as the lumenal binding protein BiP, assist newly synthesized polypeptides in acquiring their correct conformation. Proteins that fail to attain the correct three-dimensional structure are eventually degraded by a mechanism that does not involve the Golgi complex–mediated route to the vacuole. Alternatively, some proteins that are not properly folded in the ER are delivered back to the cytosol by reverse translocation across the ER.

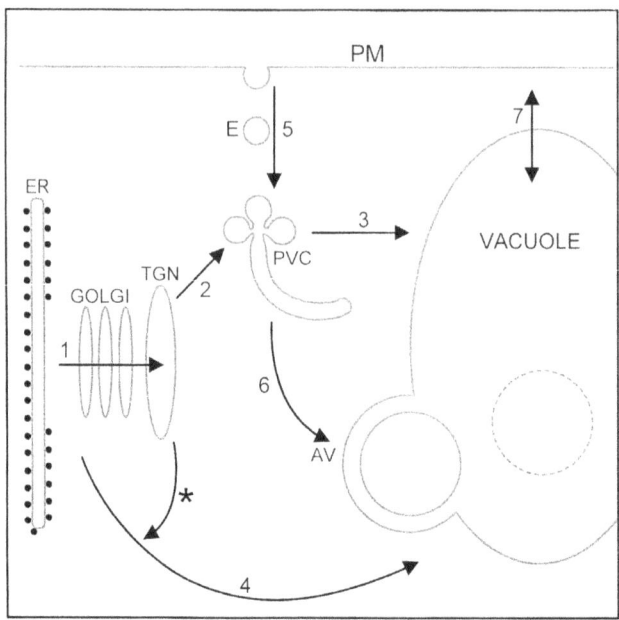

A Working Model for Transport Pathways in the Vacuolar Apparatus.

Seven basic pathways are used for the biogenesis, maintenance, and supplying of vacuoles. Pathway 1: entry and transport in the early secretory pathway (from ER to late Golgi compartments). Pathway 2: sorting of vacuolar proteins in the trans-Golgi network (TGN) to a pre/provacuolar compartment (PVC) via an early biosynthetic vacuolar pathway. Pathway 3: transport from PVC to vacuole via the late biosynthetic vacuolar pathway. Pathway 4: transport from early secretory steps (ER to Golgi complex; pathway 1) to the vacuole via an alternative route with possible material accretion from Golgi (indicated by the asterisk). Pathway 5: endocytotic pathway from the cell surface to the vacuole via endosomes. Pathway 6: cytoplasm to vacuole through autophagy by degradative or biosynthetic pathways. Pathway 7: transport of ions and solutes across the tonoplast. AV, autophagic vacuole; E, early endosome; ER, endoplasmic reticulum; PVC, pre/provacuolar compartment; TGN, trans-Golgi network.

Secretory proteins that are inserted into or translocated across the ER membrane can contain sorting signals required for their targeting to and/or retention in almost any of the compartments along the secretory pathway. For some proteins, the target organelle is the ER itself, and these proteins are not transported further. All other proteins competent for transport along the secretory pathway are carried to the Golgi complex via a still elusive vesiculo-tubular intermediate compartment. Indeed, tubular continuities have been shown to form direct linkages between the ER and the Golgi complex. Consequently, tubular transport might occur in a direction tangential, rather than perpendicular, to the Golgi stacks, in a manner that differs, therefore, from that usually assumed to operate in animal and fungal cells.

The Golgi complex has a pivotal role in the secretory pathway. In plant cells, it consists of a set of dispersed units (dictyosomes) surrounded by a proteinaceous matrix. Like its counterpart in animal cells, each morphological Golgi unit in the plant cell includes a Golgi stack and a trans-Golgi network. The Golgi stacks consist of three discrete groups of cisternae (cis, medial, and trans) that can be defined by their distinct morphologies and by their cytochemical and biochemical properties. Covalent and conformational modifications of newly synthesized secretory proteins, which begin in the ER, are continued in the Golgi complex and post-Golgi compartments. As they are being processed, vacuolar proteins transit through the early stages of the secretory pathway together with proteins that are destined to be exported into the extracellular medium or delivered to the plasma membrane.

Late Secretory Pathway—from TGN to Prevacuoles

The TGN is a major branch point in the secretory pathway and is the site of multiple sorting events that separate proteins destined for exocytotic egress from those progressing to the vacuole. The TGN varies in size according to the specific function of the cell. Under the hypothesis that biogenetic and trafficking processes are modulated in response to specific cell requirements, comprehensive morphological studies have been performed in actively vacuolating cells. The processes involved in the formation of vacuoles and their partitioning during mitosis, for example, are conveniently studied in the differentiating cells of the root meristem. Figure shows the partitioning of mitotic provacuole clusters into daughter cells.

In cells in which new vacuoles are being formed, the TGN consists of a twisted, polygonal meshwork of smooth-surfaced anastomosing tubules extending from a central disk-shaped cisternal cavity facing the Golgi stack. Via lateral linkages, a single TGN might be shared by several Golgi units. Clathrin-coated blebs and local swellings containing internal vesicles can be observed along the tubules, and numerous vesicles budding from the TGN mediate the transport of biomolecules to the vacuole.

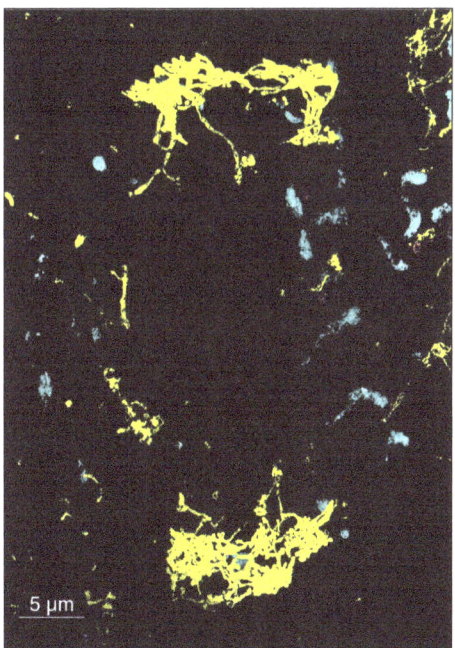

Partitioning of the Vacuolar Apparatus during Mitosis.

Distribution of mitotic provacuole clusters in vacuolating cells from the root meristem of horseradish. The vacuolar apparatus was selectively labeled by the zinc iodide–osmium reaction. The specimen (3 μm thick) was examined without counterstain at 2.5 MV with a very high voltage (3 MV) electron microscope. Images were processed using Photoshop software (Adobe, San Jose, CA). Provacuoles are shown in yellow, Golgi complexes in red, and mitochondria and plastids in blue.

The Prevacuolar Compartment

The TGN-derived vesicles on the vacuolar pathway form an intermediate compartment between the late trans-Golgi sorting site and the vacuole. These vesicles have been collectively referred to as provacuoles because they act ontogenetically as the immediate progenitors of the vacuole. They also mediate transport between the ER/Golgi complex and the vacuole and thus take functional precedence in the path to the vacuole. On account of this succession, they can be said to act as a physiological prevacuolar compartment (PVC) for cargo proteins en route to the vacuole.

Nascent provacuoles, budding from nodes of the TGN meshwork, have an average diameter (~100 nm) distinctly larger than the diameter of the TGN tubules (~15 nm). Rather quickly, the vesicular provacuoles extend into tubular provacuoles having roughly the same bore (100 nm) as the vesicles from which they derive. Their lumen is filled with vesicles that are presumably derived from microinvagination of their membranes (F. Marty, unpublished observations).

The extensive tubular provacuoles in vacuolating cells may be an enhanced version of the ubiquitous PVC described in mammalian cells and yeast. The membrane proliferation results from a dynamic effect that would occur either if membrane flow out of the provacuole were slowed down or if the membrane input from the TGN and/or the endocytotic tributary were increased. Furthermore, the provacuolar compartment might be a critical junction in post-Golgi trafficking at which the endocytotic and vacuolar biogenetic pathways converge.

Autophagy and Vacuolation

As revealed by three-dimensional high-voltage electron microscopy, the formation of autophagic vacuoles begins with a striking sequence of provacuole tubulation that proceeds to enclose discrete volumes of cytoplasmic material. Figure represents this sequence of events, whereby tubular provacuoles produce digitate extensions that form cagelike traps so as to sequester portions of cytoplasm. Adjacent bars of the cage then fuse in a zipperlike fashion and, through transient palmar connections, build a continuous and tight cavity around the segregated portion of cytoplasm. Sections through these ball-shaped structures are recognized as early autophagosomes (i.e., a cytoplasmic area encircled by a narrow ringlike cavity bounded by inner and outer membranes).

Cytochemical studies show that the TGN, provacuoles, and autophagosomes are acidic and contain lysosomal acid hydrolases. The cytoplasm in the autophagosome is degraded after it has been totally closed off. It is speculated that the digestive enzymes are released from the surrounding cavity as the inner boundary membrane deteriorates. Upon completion of the digestive process, a typical vacuole is formed. The outer membrane, which remains impermeable to hydrolytic enzymes, confines digestive activities within the forming vacuole and becomes the tonoplast. Newly formed vacuoles can then fuse together to produce a few large vacuoles. Ultimately, facilitated transport of water through the tonoplast, mediated by γ-TIP aquaporins, results in rapid vacuole enlargement.

Sequential stages of cytoplasmic confinement by provacuoles involved in cellular autophagy. Tubular provacuoles (1 and 2) form cagelike traps (2 to 4) enclosing portions of cytoplasm of a cell from the root meristem of Euphorbia characias. Adjacent bars of the cages then fuse to build a continuous and tight envelope (central structure) around segregated portions of cytoplasm. Samples were processed as described in the legend to figure.

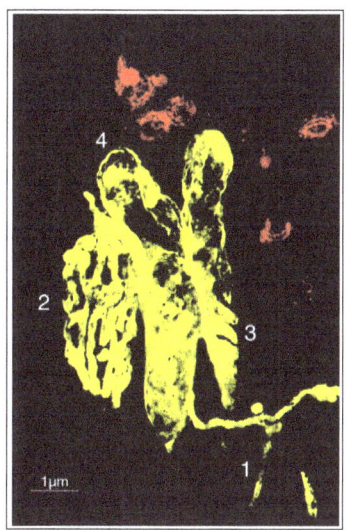

Autophagic Activity of Provacuoles.

Starvation-induced Autophagy

In response to starvation, autophagy is reinitiated in cells that are already vacuolated. The autophagic pathway is activated, for example, after sucrose deprivation of suspension-cultured cells. Portions of peripheral cytoplasm are first sequestered in double membrane–bounded envelopes (through the process described above) and then eventually digested. The small vacuoles, thus newly formed in the cytoplasm, are finally incorporated into the central vacuole. It has been shown that the induction of cellular autophagy is controlled by the supply of mitochondria with respiratory substrate and not by the decrease in the concentration of sucrose and hexose phosphates. Formation of autophagic vacuoles has been correlated with an increase in the rates of intracellular proteolysis and a massive breakdown of membrane polar lipids.

As a degradative pathway, autophagy plays a central role in protein and organelle turnover. It has been implicated in vacuolation and cell differentiation, and it is critical for survival during stress conditions such as nutrient deprivation. It can also be exploited for biosynthetic purposes as a cytoplasm-to-vacuole targeting pathway, as occurs in yeast, and with regard to supplying PSVs.

Vacuolar Sorting of Storage Proteins

Specialized cells in seeds and vegetative organs accumulate proteins that function primarily as reserves of amino acids. The most common storage proteins are the globulins, which are found in embryos, and the prolamins, which are unique to cereal endosperms. Most storage proteins, including globulins and some prolamins, have been shown to be transported to vacuoles via the Golgi complex. However, studies on the assembly and transport of seed storage proteins in legumes and cereals have shown that reserve proteins can be sorted at diverse exit sites along the vacuolar pathway. As a result, proteins are stored in a variety of compartments specific to the plant species, tissue, stage of cell differentiation, and protein category.

Golgi-dependent and Golgi-independent Routes to PSVs

Pulse-labeling experiments, morphological and immunocytochemical studies, and biochemical analyses have provided compelling evidence for a Golgi-mediated route to the PSV in legumes and other dicots. Storage proteins are synthesized as precursors that are cotranslationally transferred into the lumen of the rough ER and transported via the Golgi apparatus into specialized vacuoles where proteolytic processing is usually needed to promote their stable storage.

Whereas protein storage deposits are seldom observed in the lumen of the rough ER at the early stage of the transport pathway, condensed storage proteins are commonly detected in smooth-surfaced vesicles, ~100 nm in diameter, in association with the cis-, medial-, or trans-cisternae of the Golgi complex. Furthermore, three different types of vesicle are commonly found in close proximity to the Golgi area: vesicles that carry storage proteins, exocytotic vesicles containing cell wall polymers, and clathrin-coated vesicles (CCVs). The existence of two different exit sites for vacuolar proteins at the Golgi complex and the utilization of "alternative" secretory vesicles suggest further variations to vacuole function. According to current views, vesicles containing storage proteins originate at the cis-Golgi cisternae, and proteins undergo maturation processes as they progress through the Golgi stacks up to the TGN, where they are sorted to the storage vacuoles. At the exit site, CCVs were found to bud off from the vesicles containing storage proteins. Subsequently, only the clathrin-free vesicles, but not the CCVs, are involved in the transport of soluble storage proteins to the vacuole.

A different pathway recently has been suggested in cells from maturing seeds of pumpkin and castor bean. In these cells, proglobulin and pro2S albumin were shown to be transferred from the rough ER to the PSV via large vesicles (200 to 400 nm in diameter). These large precursor-accumulating vesicles are distinct from the Golgi-derived vesicles but similar to the late protein bodies described in pea cotyledons. It was suggested that the core of storage proproteins contained in these large vesicles might derive directly from protein aggregates that are formed in the ER; they accumulate proprotein precursors and ER-resident proteins such as BiP but not mature products. In maturing pumpkin cotyledons, where the vast majority of storage proteins are not glycosylated, the precursor-accumulating vesicles bypass the Golgi apparatus such that their transport is not inhibited by the carboxylic ionophore monensin.

In contrast to pumpkin seeds, castor bean seeds contain storage glycoproteins with complex glycans. Their processing occurs in the Golgi complex. The Golgi-processed glycoproteins are subsequently incorporated into the ER-derived precursor-accumulating vesicles at the periphery of the core aggregates. Storage glycoproteins, together with other storage proteins, are ultimately transported by the mature vesicle as far as the PSV. However, the final steps of the transport pathway to the storage vacuole are still unknown. It was suggested that the incorporation of the precursors into PSVs could occur by membrane fusion or by autophagic engulfment of the vesicle into the vacuole.

Autophagy and PSVs

Developing legume cotyledons comprise a model system to study both the ontogenesis of the PSV and the intracellular transport of vacuolar reserve proteins. In the parenchyma cells of maturing legume cotyledons, the very few large vegetative vacuoles become replaced by numerous PSVs. Ultrastructural studies indicate that the preexisting vegetative vacuoles of immature parenchyma cells are trapped by a newly developing smooth tubulo-cisternal membrane system that already contains storage proteins. However, the origin of this new membrane system is not clearly understood. The trapped vegetative vacuoles disappear as the novel storage vacuoles gradually fill up with storage proteins. During the process, the storage proteins aggregate as individual clumps against the tonoplast and cause it to protrude into the cytoplasm. By a budding process, the protruding protein masses, still surrounded by the tonoplast, become independent small storage vacuoles (membrane bounded "protein bodies") dispersed in the cytoplasm.

At later stages of cotyledon maturation, the budding process stops, and the main original storage vacuole, which continues to accumulate reserve proteins, transforms into a distinct category of large storage vacuole. A third type of storage protein reservoir is formed in the cells at the middle to late stages of seed maturation, before storage protein synthesis ceases. Storage proteins accumulate in smooth-surfaced cisternae and channels with terminal dilations. These swellings may detach and become independent spherical bodies without cisternal connections.

Finally, in germinating legume seedlings, PSVs are replaced by a vegetative vacuole through yet another type of developmentally regulated sequestration and disposal of organelles. Local invaginations of the tonoplast and engulfment of cytoplasmic fragments, subsequently degraded in the PSV, have been described.

Storage Proteins in Cereals

Cereal grains differ from legume seeds by accumulating the alcohol-soluble prolamins as storage proteins in endosperm cells. Cereal prolamins, like legume globulins, are cotranslationally loaded into the lumen of the ER. In many cereals, including maize, rice, and sorghum, prolamins form dense, insoluble accretions, which are retained within the lumen of the ER and, as in the case of the legumes, termed PBs. In developing endosperm cells, PBs become enlarged as newly synthesized prolamins are acquired and assembled with the aid of protein disulfide isomerase and molecular chaperones such as BiP.

Prolamins of other cereals, including wheat, barley, and oat, on the other hand, accumulate in vacuoles together with globulins. Globulins are transported along the anterograde pathway via the Golgi complex to the vacuolar compartment, whereas prolamin PBs are incorporated into the vacuole by an autophagic process.

Several cytological observations have suggested that rather similar autophagic mechanisms might operate when transgenes encoding storage proteins from cereals are expressed in vegetative tobacco cells. The transgene products form accretions in the ER, as in many storage cells in cereals, but the ER membrane–bounded PBs are subsequently captured by an autophagic process and delivered to the vegetative vacuole, where they are eventually proteolytically degraded. Interestingly, somewhat similar steps could be detected during the transport of storage proteins to storage vacuoles by large precursor-accumulating vesicles in normally developing cells. These results suggest that the cellular machinery of autophagy can be used for delivering cytosolic proteins and early membrane-bounded PBs to the vacuole, thus defining a biosynthetic cytoplasm-to-vacuole targeting pathway as occurs in yeast. The ontogeny of the compartments specialized in protein storage is thus diverse, and not all stores are (ontogenetically) homologous, although all belong to the vacuolar apparatus of plant cells.

Endocytosis

Endocytosis is defined as the uptake of extracellular and plasma membrane materials from the cell surface into the cell. Endocytosis has been characterized morphologically in plant cells in which both fluid-phase uptake and receptor-mediated internalizations have been visualized. Two distinct routes of internalization by clathrin-mediated endocytosis have been suggested to operate in plant cells: from the plasma membrane to an endosomal compartment, including the partially coated reticulum, multivesicular bodies, TGN, and the PVC; and from the plasma membrane to the PVC and the vacuoles. Novel intermediary structures arising from plasma membrane internalization have also been described as part of a compensatory recycling mechanism in actively secreting cells. Rapid retrieval of plasma membrane to the cell interior, together with a fluid phase internalization of extracellular material, occurs in water-stressed cells.

Morphological studies of vacuolating cells by electron microscopy suggest that the endocytotic and biosynthetic vacuolar pathways converge at the provacuolar compartment before nascent autophagic vacuoles are formed (F. Marty, unpublished observations). The convergence point(s) between these pathways in already vacuolated cells is unknown, but it seems reasonable to hypothesize that the juncture could be at the prevacuolar compartment. Endocytotic vesicles and endosomes belong to the vacuolar apparatus, but their direct contribution to the formation of the vacuole remains uncertain. Whereas the vesicle-mediated internalization of plasma membrane has been documented in plant cells, the routes involved need to be precisely mapped by reliable tracers. A potential candidate is Tlg1p, a protein functionally homologous to the t-SNARE localized on a putative early endosome in yeast.

Vacuolar Sorting Signals

Vacuolar soluble proteins and membrane proteins alike travel through the early stages of the secretory pathway. Most probably, they are sorted away from proteins destined for delivery to the cell surface at the exit of the Golgi complex. Soluble proteins therefore require a sorting signal to tag them for vacuolar delivery after their egress from the Golgi complex; indeed, in the absence of such informational tags, they are secreted to the extracellular space. Three types of vacuolar targeting signals have been described. Some vacuolar proteins (e.g., sporamin and aleurain) contain an N-terminal propeptide (NTPP) as a targeting determinant; others (e.g., barley lectin, phaseolin, tobacco chitinase, and Brazil nut 2S albumin) contain a C-terminal propeptide (CTPP), whereas some vacuolar proteins (e.g., phytohemagglutinin and legumin) contain a targeting signal in an exposed region of the mature protein.

NTPP Signals

The targeting determinants characterized in NTPPs from the barley cysteine protease aleurain and from sweet potato sporamin contain a conserved Asn-Pro-Ile-Arg amino acid sequence. This motif in the NTPP is necessary and sufficient for the sorting of the sporamin precursor to the vacuole. Sporamin is delivered to the sink vacuole in cells from the tuberous roots of the sweet potato, whereas aleurain is sorted to a lytic compartment distinct from the PSV.

CTPP Signals

By contrast to NTPP signals, a vacuolar sorting consensus sequence has not been identified in CTPP targeting domains. Nevertheless, the CTPP was shown to be necessary and sufficient for the targeting of barley lectin to the vacuole. The N-linked glycan of the CTPP in barley lectin is not necessary for sorting, although it modulates the rate of processing of the propeptide. Hydrophobic residues in the CTPP are important for the targeting of barley lectin. Similar mutagenesis analyses have been performed to characterize the targeting signal of tobacco chitinase. CTPPs from vacuolar proteins differ in length, and it was recently shown that a short CTPP from phaseolin contains information necessary for interactions with the vacuolar sorting machinery in a saturable manner.

The barley lectin, phaseolin, and Brazil nut 2S albumin accumulate in PSVs, whereas tobacco chitinase is delivered to vacuoles of vegetative cells. Results indicate that more than one sorting mechanism might exploit the CTPP targeting signal and that transport of CTPP-containing proteins from the Golgi complex to the vacuoles involves more than one pathway. Both CTPP- and NTPP-mediated vacuolar delivery also involve alternative structures and mechanisms, although NTPP and CTPP were found to be functionally interchangeable in directing proteins to the vacuole.

Internal Signals

Other plant vacuolar proteins are synthesized without a cleavable vacuolar-targeting signal. Studies on phytohemagglutinin (PHA) from Phaseolus vulgaris and legumin from Vicia faba have demonstrated targeting information in exposed regions of the mature proteins, which are deposited in the PSVs of the reserve parenchyma cells of cotyledons. Strikingly, soluble proteins, such as PHA, and proteinase inhibitors, which are usually vacuolar, occasionally have been detected in the extracellular matrix, suggesting that the vacuolar targeting signals might not be recognized in all cells. Moreover, recent work on suspension-cultured cells showed that some soluble, fully processed, vacuolar hydrolases can be excreted into the medium under hormonal control. The exocytotic pathway for these "vacuolar" proteins would lead from either the vacuole or the PVC situated downstream of the last processing step.

Although short amino acid sequences of plant vacuolar proteins are sufficient to sort nonvacuolar proteins to the vacuole in yeast, plant proteins are sorted to the yeast vacuole by signals different from those recognized by plants, suggesting that the transport machinery is at least partially different between yeast and plants.

Vacuolar membrane and intravacuolar soluble proteins are targeted to vacuoles by different mechanisms. Pulse–chase experiments and pharmacological studies on protoplasts from transgenic tobacco plants suggest that soluble proteins such as PHA and integral membrane proteins such as α-TIP reach the same destination by traveling through different paths.

Signals in TIPs

Transport pathways for integral membrane proteins of the tonoplast have been investigated. The vacuolar membrane α-TIP and γ-TIP are polytopic integral membrane proteins, with six membrane-spanning domains and both N and C termini located in the cytoplasm. In an early analysis of the targeting information contained in α-TIP, it was found that a polypeptide segment comprising the sixth membrane-spanning domain and the adjacent C-terminal, cytoplasmic tail of α-TIP is sufficient to target a nonvacuolar reporter protein to the tonoplast. In addition, the C-terminal cytoplasmic tail was not found necessary for the targeting of α-TIP in the same stably transformed tobacco cells.

More recently, the trafficking of a chimeric integral membrane reporter protein was analyzed in tobacco protoplasts. It was found that the transmembrane domain of the plant vacuolar sorting receptor BP-80 directs the reporter protein via the Golgi complex to the prevacuolar compartment, and attaching the C-terminal cytoplasmic tail of γ-TIP did not alter this traffic. By contrast, attaching the C-terminal cytoplasmic tail of α-TIP prevented traffic of the reporter protein through the Golgi complex but caused it to be localized to vacuoles. It was thus concluded that there are two separate pathways to vacuoles for membrane proteins: a direct pathway followed by α-TIP from the ER to PSVs, and a separate pathway followed by γ-TIP via the Golgi complex and PVC to the vegetative lytic vacuole.

Vacuolar Sorting Receptors

Soluble vacuolar proteins are diverted away from the exocytotic pathway through a receptor-mediated process that leads to their delivery to the vacuole. Two independent approaches resulted in the identification of plant vacuolar sorting receptors.

It was initially hypothesized that the Asn-Pro-Ile-Arg motif conserved in the NTPP vacuole-targeting determinant of aleurain and sporamin, two unrelated proteins, was likely to be recognized by a sorting receptor. Indeed, a protein of 80 kD, called BP-80, has been affinity purified from a lysate of CCVs from pea. It possesses all the features expected of a vacuolar sorting receptor. It is a type I integral membrane protein that is localized in the Golgi complex and in small vacuolar structures. These vacuolar structures are distinct from both α-TIP and γ-TIP vacuoles but are possibly analogous to prevacuoles. Several homologs have been cloned, and the sequences appear to be highly conserved in monocotyledonous and dicotyledonous plants.

An alternative approach led to the identification of an Arabidopsis receptor-like protein called AtELP (for Arabidopsis thaliana epidermal growth factor–like protein). This second approach was based on the use of known functional motifs present in many of the receptor proteins involved in clathrin-dependent intracellular protein sorting in mammalian and yeast cells. AtELP shares many common features with mammalian and yeast transmembrane cargo receptors. It is capable of in vitro interaction with the proteins of the TGN-specific AP-1 adaptor complex from mammals. It is located at the TGN, in CCVs, and on the PVC in the root cells of Arabidopsis. AtELP is highly homologous to BP-80, suggesting that it also may play a role in targeting proteins to the vacuole. Mechanisms recognizing the C-terminal or internal vacuolar sorting signals of soluble proteins have not been elucidated, and the identification of receptor-mediated pathways for membrane proteins is still in debate.

In addition to sorting receptors, other components of the vacuolar targeting machinery are being identified in plants. An interesting example is a V-ATPase activity associated with the Golgi complex, distinct from that of the tonoplast V-ATPase, and which is necessary for the efficient targeting of soluble proteins to the vacuole.

Trafficking Steps and Snare Components

Transport of soluble and membrane proteins in the secretory pathway is known to be mediated by the budding and fusion of transport vesicles and, in certain cell types or physiological situations, by cisternal progression and direct tubular linkages between different compartments.

As an early step in vesicular transport, budding involves coat proteins that assemble from the cytosol. CCVs, COPI and COPII-like vesicles, and "dense" vesicles have been described in plant cells. Available data indicate that a considerable homology between coat proteins in plant, yeast, and animal cells exists, although we still know little of the molecular organization of transport vesicles in plants.

Docking and fusion steps are thought to be mainly regulated by integral membrane receptors, termed SNAREs (for soluble N-ethylmaleimide–sensitive factor attachment protein receptors). According to the prevalent model, SNAREs on vesicles (v-SNAREs) interact with cognate SNAREs on the target membranes (t-SNAREs). The soluble proteins NSF (for N-ethylmaleimide–sensitive factor) and α-SNAP (for soluble NSF attachment protein) then bind the v-SNARE/t-SNARE complex, and a rearrangement triggered by ATP hydrolysis finally promotes membrane fusion. The diversity and specificity of vesicle transport routes correlate with the complexity of traffic effectors, which include Rab proteins, Rab-binding molecules, Ca^{2+}, and components of the cytoskeleton. Many lines of investigation suggest that the mechanisms of vesicular budding, docking, and fusion are conserved across species and subcellular compartments.

A growing number of proteins functionally homologous (orthologs) to the SNAREs characterized in yeast and mammalian cells is being identified in plant cells. Initial results show that the sorting mechanism for soluble proteins to the plant vacuole agrees well with the SNARE model. The putative plant vacuolar receptor AtELP is able to recruit the adaptor complex 1 (AP-1) present at the TGN. As a consequence, the AtELP receptor appears to be included in TGN-derived CCVs. These vesicles carry the vacuolar cargo together with its receptor to the prevacuolar compartment, where the receptor (AtELP) and the prevacuole-specific t-SNARE are colocalized. The vacuolar t-SNARE AtVam3p is used downstream in the late vacuolar pathway. However, its function in homotypic (vacuole–vacuole) or heterotypic (prevacuole–vacuole) fusions or in autophagy is still being debated.

Compelling microscopic evidence is also suggestive of transient tubular continuities between compartments of the vacuolar pathway in particular cell types or physiological conditions. Indeed, tubular continuities between the ER and the Golgi complex, between cisternae from the same Golgi stack, between TGN units from adjacent Golgi stacks, between the TGN and the pre/provacuolar compartment, and between provacuoles and autophagic vacuoles have been described. Such interconnections are consistent with an intracellular transport by cisternal progression and maturation. Vesicular and nonvesicular transport mechanisms, it should be stressed, are not mutually exclusive.

Tonoplast Functions

The vacuole plays an important role in the homeostasis of the plant cell. It is involved in the control of cell volume and cell turgor. The regulation of cytoplasmic ions and pH of the storage of amino acids, sugars, and CO_2 and the sequestration of toxic ions and xenobiotics. These activities are driven by specific proteins present in the tonoplast and indicated in figure.

According to the chemiosmotic model for energy-dependent solute transport, the proton-motive force generated by either the V-ATPase or the H^+-translocating inorganic pyrophosphatase

(V-PPase) can be used to drive secondary solute transports. Movement of ions and water down their thermodynamic potentials is achieved by specific ion channels and water channels (aquaporins). The resulting ion, water, and metabolite fluxes across the vacuolar membrane are crucial to the diverse functions of the vacuole in plant cells, such as cell enlargement and plant growth, signal transduction, protoplasmic homeostasis, and regulation of metabolic pathways.

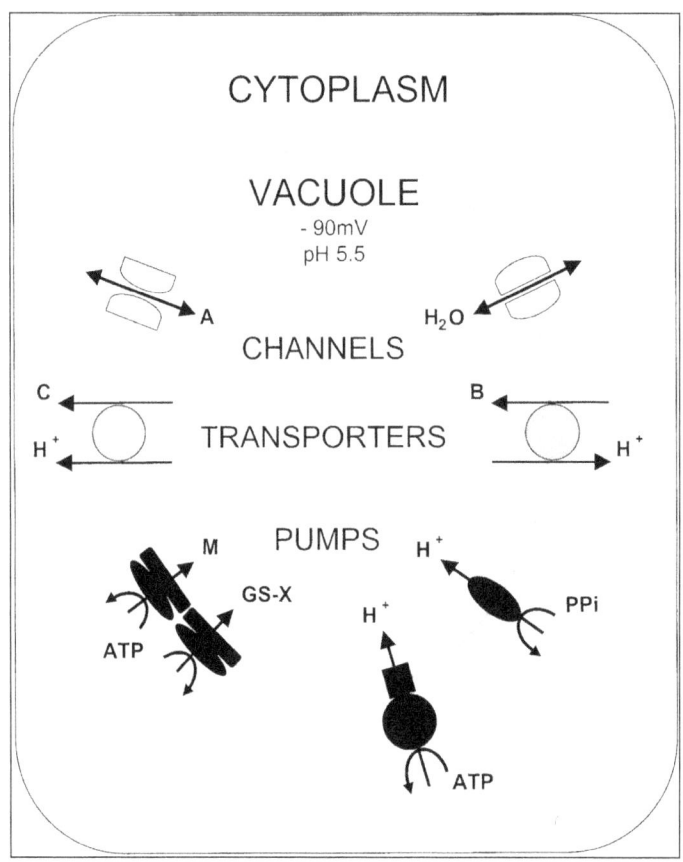

Model of ABC Transporters, H^+ Primary Pumps, H^+-Coupled Transporters, and Channels in a Simplified Tonoplast.

Glutathione S-conjugate (GS-X) and metabolite (M) transport is achieved by an ABC transporter(s). An electrogenic H^+-ATPase (V-type) and an H^+-PPase acidify the vacuole. The proton motive force provides energy for uptake and release of solutes (i.e., cations, anions, and organic solutes, denoted A, B, or C indiscriminately here) across the tonoplast through transporters and channels. Water channels (aquaporins) facilitate the passive exchange of water.

Recent studies have demonstrated the existence of a group of organic solute transporters, belonging to the ABC superfamily, that are directly energized by MgATP. These pumps are competent in the transport of a broad range of substances, including sugars, peptides, alkaloids, and inorganic anions. Belonging to the ABC family, the multidrug resistance–associated proteins (MRPs) identified in plants are considered to participate in the transport of exogenous and endogenous amphipathic anions and glutathionated compounds from the cytoplasm to the vacuole. They function in herbicide detoxification, cell pigmentation, storage of antimicrobial compounds, and alleviation of oxidative damage. A role for plant MRPs is also suspected in channel regulation and transport of heavy metal chelates.

References

- Cell-wall-plant-anatomy: britannica.com, Retrieved 14 May, 2019

- Middle-Lamella: plant-biology, Retrieved 15 April, 2019

- Protoplasm: byjus.com, Retrieved 06 May, 2019

- Cytoplasm-function-and-facts-13714432: sciencing.com, Retrieved 03 January, 2019

- Chloroplast: newworldencyclopedia.org, Retrieved 20 February, 2019

- Amyloplast-definition-4142136: thoughtco.com, Retrieved 12 March, 2019

- Endoplasmic-reticulum: biologydictionary.net, Retrieved 29 August, 2019

- Endoplasmic-reticulum-rough-and-smooth, softcell-e-learning: bscb.org, Retrieved 15 June, 2019

4

The Plant Hormones

Plant hormones are molecules produced within the plants in low concentrations. A few of these hormones are auxin, gibberlin, cytokinin, jasmonate, karrikin, florigen, ethylene, peptide, strigolactone, etc. All these different plant hormones have been carefully analyzed in this chapter.

Plant hormones are chemicals plants use for communication, coordination, and development between their many cells. Like animals, plants rely on these chemical signals to direct the expression of DNA and the operations of the cell. Plant hormones are natural substances which control many aspects of plant development. They control everything from the length between nodes on the branches to the programmed death, or senescence seen in many annual plants.

There are various classes of plant hormone, each which controls various aspects of plant development. There are also several other recently recognized plant hormones. Remember that these are general categories, and that individual species may have developed novel uses for various hormones.

Auxins

An auxin is a plant hormone derived from the amino acid tryptophan. An auxin may be one of many molecules, but all auxin molecules are involved in some sort of cellular regulation. Auxin molecules are one of five major types of plant hormone. The other major groups are the gibberellins, cytokinins, ethylene, and abscisic acid. Auxin was the first of these groups to be identified, and was chemically isolated in the 1930's.

The most widespread auxin is indoleacetic acid, or simply IAA. IAA is an auxin which is very important in the growth and development of plant tissues. In studying auxin molecules, scientist have been able to recreate similar structures, called synthetic growth regulators. These "fake" auxins also stimulate growth in plants and have been used in many agricultural and commercial applications.

Auxin Function

The auxin group of hormones has a wide range of uses in a plant. Auxin molecules are found in all tissues in a plant. However, they tend to be concentrated in the meristems, growth centers which are at the forefront of growth. These centers release auxin molecules, which are then distributed towards the roots. In this way, the plant can coordinate its size, and the growth and development of different tissues based on the gradient of the auxin concentration.

Auxin affects many different cellular processes. At the molecular level, auxin molecules can affect cytoplasmic streaming, the movement of fluids within a cell, and even the activity of various enzymes. This gives auxin direct control over the growth, development, and proliferation of individual cells within the plant. The auxin gradient directly affects processes such as flower initiation, fruit development, and even tuber and bulb formation. Even on a daily basis, auxin levels affect processes such as phototropism, which allows the plant to follow the sun and gain the most energy. The auxin controls this process by concentrating in the side of the plant away from the sun. This causes changes in the cells, which bend the plant toward the light. This can be seen in the image below:

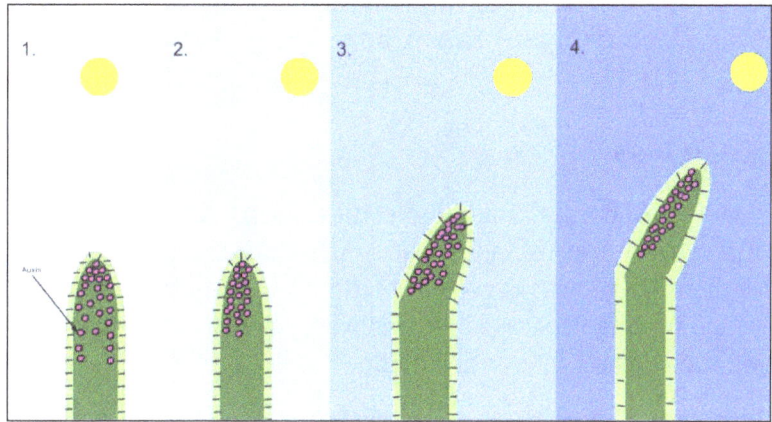

Another important feature which auxin gradients provide many plants is apical dominance. Apical dominance is formed when a single meristem is growing faster and more efficiently. Eventually, the auxin released from this meristem inhibits any new shoots from budding off below it. If the stem is cut off, many new shoot will erupt below the stem, as the auxin gradient has been disrupted and the system must create a new leading shoot. The auxin gradient, when established, determines how fast internodes grow, which determines the height of the plant. When discussing the function of the auxin molecules in a plant, it is almost easier to discuss the things they do not control.

Some scientist have even discussed the polar-auxin transport system as a plant-like take on a nervous system. The way the auxin molecules move from cell to cell is very similar to how a nerve signal is sent across an animal's body. The auxin molecule affects various tissues, and is usually converted into another auxin. A "return signal" can then be generated. In this way, using the many different versions of auxin and the other plant hormones, a plant could have a feasibly robust nervous system for responding to external stimuli.

Auxin Structure

Native auxin molecules are normally derived from the amino acid tryptophan. This amino acid has a six-sided carbon ring, attached to a 5-sided ring containing carbon. This 5-sided ring has a group attached. The only difference between most auxin molecules and tryptophan is what is attached to this ring. The common auxin IAA can be seen below:

To create this molecule, two enzymes are needed to act on tryptophan. First, an amino-transferase removes a nitrogen and a hydrogen from the side-chain attached to the 5-sided ring. Then, a decarboxylase enzyme removes the carboxyl group, leaving only COOH. A chloride ion attaches to the six-sided ring, and IAA is born. Most auxins are some derivation of this molecule.

Synthetic Auxin Analogs

After studying the structure of natural auxin molecules, scientist were easily able to produce molecules which were similar to natural auxins. These synthetic auxin analogs have many applications. They can be used to encourage growth in certain plants. Synthetic auxin treatment is used on many plant cuttings, to induce rooting processes. In this way, scientist can make plant clones by taking cuttings, and growing the cuttings into entire plants.

1-Naphthaleneacetic acid (NAA) is a coming rooting chemical, and a synthetic auxin. This fake auxin is marketing to regular gardeners. While there are some safety and handling concerns, fake auxin molecules have been used since the 1940's to stimulate the growth of cuttings. Scientist also found that auxin molecules could have anti-growth properties as well.

The synthetic auxin 2,4-D (2,4-Dichlorophenoxyacetic acid), is a common weed killer. The auxin-like molecule affects only broadleaf weed species. This means it can be applied around lawn, grassland, and other landscape plants without affecting them. However, in the broadleaf plants it causes rapid growth in all the wrong places. The plants quickly die off. There are many other synthetic auxin compounds, which have a variety of marketed uses.

Gibberellins

Gibberellins (GAs) are plant hormones that regulate various developmental processes, including stem elongation, germination, dormancy, flowering, flower development, and leaf and fruit senescence. GAs are one of the longest-known classes of plant hormone. It is thought that the selective breeding (albeit unconscious) of crop strains that were deficient in GA synthesis was one of the key drivers of the "green revolution" in the 1960s, a revolution that is credited to have saved over a billion lives worldwide.

Chemistry

All known gibberellins are diterpenoid acids that are synthesized by the terpenoid pathway in plastids and then modified in the endoplasmic reticulum and cytosol until they reach their biologically-active form. All gibberellins are derived via the ent-gibberellane skeleton, but are synthesised via ent-kaurene. The gibberellins are named GA1 through GAn in order of discovery. Gibberellic acid, which was the first gibberellin to be structurally characterized, is GA3.

As of 2003, there were 126 GAs identified from plants, fungi, and bacteria.

Gibberellins are tetracyclic diterpene acids. There are two classes based on the presence of either 19 or 20 carbons. The 19-carbon gibberellins, such as gibberellic acid, have lost carbon 20 and, in place, possess a five-member lactone bridge that links carbons 4 and 10. The 19-carbon forms are, in general, the biologically active forms of gibberellins. Hydroxylation also has a great effect on the biological activity of the gibberellin. In general, the most biologically active compounds are dihydroxylated gibberellins, which possess hydroxyl groups on both carbon 3 and carbon 13. Gibberellic acid is a dihydroxylated gibberellin.

Bioactive GAs

The bioactive GAs are GA1, GA3, GA4, and GA7. There are three common structural traits between these GAs: hydroxyl group on C-3β, a carboxyl group on C-6, and a lactone between C-4 and C-10. The 3β-hydroxyl group can be exchanged for other functional groups at C-2 and/or C-3 positions. GA5 and GA6 are examples of bioactive GAs that do not have a hydroxyl group on C-3β. The presence of GA1 in various plant species suggests that it is a common bioactive GA.

Gibberellin A1 (GA1).

Gibberellic acid (GA3).

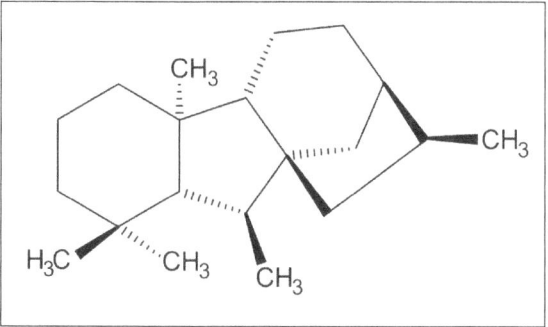

ent-Gibberellane.

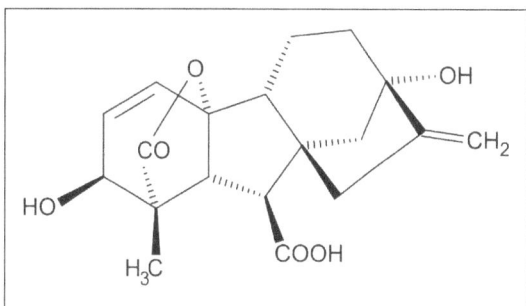

ent-Kaurene.

Biological Function

1. Shows a plant lacking gibberellins and has an internode length of "0" as well as it is a dwarf plant. 2. Shows your average plant with a moderate amount of gibberellins and an average internode length. 3.Shows a plant with a large amount of gibberellins and so has a much longer internode length because gibberellins promotes cell division in the stem.

Gibberellins are involved in the natural process of breaking dormancy and other aspects of germination. Before the photosynthetic apparatus develops sufficiently in the early stages of germination, the stored energy reserves of starch nourish the seedling. Usually in germination, the breakdown of starch to glucose in the endosperm begins shortly after the seed is exposed to water. Gibberellins in the seed embryo are believed to signal starch hydrolysis through inducing the synthesis of the enzyme α-amylase in the aleurone cells. In the model for gibberellin-induced production of α-amylase, it is demonstrated that gibberellins (denoted by GA) produced in the scutellum diffuse to the aleurone cells, where they stimulate the secretion α-amylase. α-Amylase then hydrolyses starch, which is abundant in many seeds, into glucose that can be used in cellular respiration to produce energy for the seed embryo. Studies of this process have indicated gibberellins cause higher levels of transcription of the gene coding for the α-amylase enzyme, to stimulate the synthesis of α-amylase.

Gibberellins are produced in greater mass when the plant is exposed to cold temperatures. They stimulate cell elongation, breaking and budding, seedless fruits, and seed germination. They do the last by breaking the seed's dormancy and acting as a chemical messenger. Its hormone binds to a receptor, and calcium activates the protein calmodulin, and the complex binds to DNA, producing an enzyme to stimulate growth in the embryo.

Metabolism

Biosynthesis

GAs are usually synthesized from the methylerythritol phosphate (MEP) pathway in higher plants. In this pathway, bioactive GA is produced from trans-geranylgeranyl diphosphate (GGDP). In the MEP pathway, three classes of enzymes are used to yield GA from GGDP: terpene synthases

(TPSs), cytochrome P450 monooxygenases (P450s), and 2-oxoglutarate–dependent dioxygenases (2ODDs). There are eight steps in the MEP pathway:

- GGDP is converted to ent-copalyl diphosphate (ent-CPD) by ent-copalyl diphosphate synthase.

- Etn-CDP is converted to ent-kaurene by ent-kaurene synthase.

- Ent-kaurene is converted to ent-kaurenol by ent-kaurene oxidase (KO).

- Ent-kaurenol is converted to ent-kaurenal by KO.

- Ent-kaurenal is converted to ent-kaurenoic acid by KO.

- Ent-kaurenoic acid is converted to ent-7a-hydroxykaurenoic acid by ent-kaurene acid oxidase (KAO).

- Ent-7a-hydroxykaurenoic acid is converted to GA12-aldehyde by KAO.

- GA12-aldehyde is converted to GA12 by KAO. GA12 is processed to the bioactive GA4 by oxidations on C-20 and C-3, which is accomplished by 2 soluble ODDs: GA 20-oxidase and GA 3-oxidase.

One or two genes encode the enzymes responsible for the first steps of GA biosynthesis in Arabidopsis and rice. The null alleles of the genes encoding CPS, KS, and KO result in GA-deficient Arabidopsis dwarves. Multigene families encode the 2ODDs that catalyze the formation of GA12 to bioactive GA4.

AtGA3ox1 and AtGA3ox2, two of the four genes that encode GA3ox in Arabidopsis, affect vegetative development. Environmental stimuli regulate AtGA3ox1 and AtGA3ox2 activity during seed germination. In Arabidopsis, GA20ox overexpression leads to an increase in GA concentration.

Sites of Biosynthesis

Most bioactive GAs are located in actively growing organs on plants. Both GA20ox and GA3ox genes (genes coding for GA 20-oxidase and GA 3-oxidase) and the SLENDER1 gene (a GA signal transduction gene) are found in growing organs on rice, which suggests bioactive GA synthesis occurs at their site of action in growing organs in plants. During flower development, the tapetum of anthers is believed to be a primary site of GA biosynthesis.

Differences between Biosynthesis in Fungi and Lower Plants

Arabidopsis, a plant, and Gibberella fujikuroi, a fungus, possess different GA pathways and enzymes. P450s in fungi perform functions analogous to the functions of KAOs in plants. The function of CPS and KS in plants is performed by a single enzyme, CPS/KS, in fungi. In fungi, the GA biosynthesis genes are found on one chromosome, but in plants, they are found randomly on multiple chromosomes. Plants produce low amount of GA3, therefore the GA3 is produced for industrial purposes by microorganisms. Industrially the gibberellic acid can be produced by submerged fermentation, but this process presents low yield with high production costs and hence higher sale value, nevertheless other alternative process to reduce costs of the GA3 production is solid-state fermentation (SSF) that allows the use of agro-industrial residues.

Catabolism

Several mechanisms for inactivating GAs have been identified. 2β-hydroxylation deactivates GA, and is catalyzed by GA2-oxidases (GA2oxs). Some GA2oxs use C19-GAs as substrates, and other GA2oxs use C20-GAs. Cytochrome P450 mono-oxygenase, encoded by elongated uppermost internode (eui), converts GAs into 16α,17-epoxides. Rice eui mutants amass bioactive GAs at high levels, which suggests cytochrome P450 mono-oxygenase is a main enzyme responsible for deactivation GA in rice. The Gamt1 and gamt2 genes encode enzymes that methylate the C-6 carboxyl group of GAs. In a gamt1 and gamt2 mutant, concentrations of GA is developing seeds is increased.

Homeostasis

Feedback and feedforward regulation maintains the levels of bioactive GAs in plants. Levels of AtGA20ox1 and AtGA3ox1 expression are increased in a GA deficient environment, and decreased after the addition of bioactive GAs, Conversely, expression of AtGA2ox1 and AtGA2ox2, GA deactivation genes, is increased with addition of GA.

Regulation

Regulation by other Hormones

The auxin indole-3-acetic acid (IAA) regulates concentration of GA1 in elongating internodes in peas. Removal of IAA by removal of the apical bud, the auxin source, reduces the concentration of GA1, and reintroduction of IAA reverses these effects to increase the concentration of GA1. This phenomenon has also been observed in tobacco plants. Auxin increases GA 3-oxidation and decreases GA 2-oxidation in barley. Auxin also regulates GA biosynthesis during fruit development in peas. These discoveries in different plant species suggest the auxin regulation of GA metabolism may be a universal mechanism.

Ethylene decreases the concentration of bioactive GAs.

Regulation by Environmental Factors

Recent evidence suggests fluctuations in GA concentration influence light-regulated seed germination, photomorphogenesis during de-etiolation, and photoperiod regulation of stem elongation and flowering. Microarray analysis showed about one fourth cold-responsive genes are related to GA-regulated genes, which suggests GA influences response to cold temperatures. Plants reduce growth rate when exposed to stress. A relationship between GA levels and amount of stress experienced has been suggested in barley.

Role in Seed Development

Bioactive GAs and abscisic acid levels have an inverse relationship and regulate seed development and germination. Levels of FUS3, an Arabidopsis transcription factor, are upregulated by ABA and downregulated by GA, which suggests that there is a regulation loop that establishes the balance of GA and ABA.

Signalling Mechanism

Receptor

In the early 1990's, there were several lines of evidence that suggested the existence of a GA receptor in oat seeds that was located at the plasma membrane. However despite intensive research, to date, no membrane-bound GA receptor has been isolated. This, along with the discovery of a soluble receptor, GA insensitive dwarf 1 (GID1) has led many to doubt that a membrane-bound receptor exists.

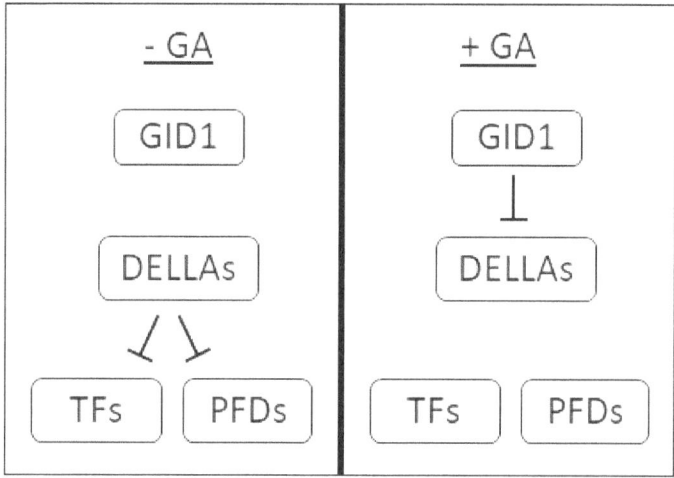

GA-GID1-DELLA signal pathway: In the absence of GA, DELLA proteins bind to and inhibit transcription factors (TFs) and prefoldins (PFDs). When GA is present, GID1 triggers the degradation of DELLAs and releases the TFs and PFDs.

GID1 was first identified in rice and in Arabidopsis there are three orthologs of GID1, AtGID1a, b, and c. GID1s have a high affinity for bioactive GAs. GA binds to a specific binding pocket on GID1; the C3-hydroxyl on GA makes contact with tyrosine-31 in the GID1 binding pocket. GA binding to GID1 causes changes in GID1 structure, causing a 'lid' on GID1 to cover the GA binding pocket. The movement of this lid results in the exposure of a surface which enables the binding of GID1 to DELLA proteins.

DELLA Proteins: Repression of a Repressor

DELLA proteins, such as SLR1 in rice or GAI and RGA in Arabidopsis are repressors of plant development. DELLAs inhibit seed germination, seed growth, flowering and GA reverses these effects. DELLA proteins are characterized by the presence of a DELLA motif (aspartate-glutamate-leucine-leucine-alanine or D-E-L-L-A in the single letter amino acid code).

When GA binds to the GID1 receptor, it enhances the interaction between GID1 and DELLA proteins, forming a GA-GID1-DELLA complex. When in the GA-GID1-DELLA complex, it is thought that DELLA proteins undergo changes in structure that enable their binding to F-box proteins

(SLY1 in Arabidopsis or GID2 in rice). F-box proteins catalyse the addition of ubiquitin to their targets. The addition of ubiquitin to DELLA proteins promotes their degradation via the 26S-proteosome. The degradation of DELLA proteins releases cells from their repressive effects.

Targets of DELLA Proteins

Transcription Factors

The first targets of DELLA proteins identified were PHYTOCHROME INTERACTING FACTORs (PIFs). PIFs are transcription factors that negatively regulate light signalling and are strong promoters of elongation growth. In the presence of GA, DELLAs are degraded and this then allows PIFs to promote elongation. It was later found that DELLAs repress a large number of other transcription factors, among which are positive regulators of auxin, brassinosteriod and ethylene signalling. DELLAs can repress transcription factors either by stopping their binding to DNA or by promoting their degradation.

Prefoldins and Microtubule Assembly

In addition to repressing transcription factors, DELLAs also bind to prefoldins (PFDs). PFDs are molecular chaperones, meaning they assist in the folding of other proteins. PFDs function in the cytosol but when DELLAs bind to PFDs, it restricts them to the nucleus. An important function of PFDs is to assist in the folding of β-tubulin. As such, in the absence of GA (when there is a high level of DELLA proteins), PDF function is reduced and there is a lower cellular pool of β-tubulin. When GA is present the DELLAs are degraded, PDFs can move to the cytosol and assist in the folding of β-tubulin. β-tubulin is a vital component of the cytoskeleton (in the form of microtubules). As such, GA allows for re-organisation of the cytoskeleton, and the elongation of cells.

Microtubules are also required for the trafficking of membrane vesicles. Membrane vesicle trafficking is needed for the correct positioning of several hormone transporters. One of the most well characterized hormone transporters are PIN proteins, which are responsible for the movement of the hormone auxin between cells. In the absence of GA, DELLA proteins reduce the levels of microtubules and thereby inhibit membrane vesicle trafficking. This reduces the level of PIN proteins at the cell membrane, and the level of auxin in the cell. GA reverses this process and allows for PIN protein trafficking to the cell membrane to enhance the level of auxin in the cell.

Cytokinins

Cytokinins (CK) are a class of plant growth substances (phytohormones) that promote cell division, or cytokinesis, in plant roots and shoots. They are involved primarily in cell growth and differentiation, but also affect apical dominance, axillary bud growth, and leaf senescence. Folke Skoog discovered their effects using coconut milk in the 1940s at the University of Wisconsin–Madison.

There are two types of cytokinins: adenine-type cytokinins represented by kinetin, zeatin, and 6-benzylaminopurine, and phenylurea-type cytokinins like diphenylurea and thidiazuron (TDZ). Most adenine-type cytokinins are synthesized in roots. Cambium and other actively dividing tissues also synthesize cytokinins. No phenylurea cytokinins have been found in plants. Cytokinins participate in local and long-distance signalling, with the same transport mechanism as purines and nucleosides. Typically, cytokinins are transported in the xylem.

Cytokinins act in concert with auxin, another plant growth hormone. The two are complementary, having generally opposite effects.

The cytokinin zeatin is named after the genus of corn, Zea, in which it was discovered.

Mode of Action

The ratio of auxin to cytokinin plays an important role in the effect of cytokinin on plant growth. Cytokinin alone has no effect on parenchyma cells. When cultured with auxin but no cytokinin, they grow large but do not divide. When cytokinin is added, the cells expand and differentiate. When cytokinin and auxin are present in equal levels, the parenchyma cells form an undifferentiated callus. More cytokinin induces growth of shoot buds, while more auxin induces root formation.

Cytokinins are involved in many plant processes, including cell division and shoot and root morphogenesis. They are known to regulate axillary bud growth and apical dominance. The "direct inhibition hypothesis" posits that these effects result from the cytokinin to auxin ratio. This theory states that auxin from apical buds travels down shoots to inhibit axillary bud growth. This promotes shoot growth, and restricts lateral branching. Cytokinin moves from the roots into the shoots, eventually signaling lateral bud growth. Simple experiments support this theory. When the apical bud is removed, the axillary buds are uninhibited, lateral growth increases, and plants become bushier. Applying auxin to the cut stem again inhibits lateral dominance.

While cytokinin action in vascular plants is described as pleiotropic, this class of plant hormones specifically induces the transition from apical growth to growth via a three-faced apical cell in moss protonema. This bud induction can be pinpointed to differentiation of a specific single cell, and thus is a very specific effect of cytokinin.

Cytokinins have been shown to slow aging of plant organs by preventing protein breakdown, activating protein synthesis, and assembling nutrients from nearby tissues. A study that regulated leaf senescence in tobacco leaves found that wild-type leaves yellowed while transgenic leaves remained mostly green. It was hypothesized that cytokinin may affect enzymes that regulate protein synthesis and degradation.

Cytokinin signaling in plants is mediated by a two-component phosphorelay. This pathway is initiated by cytokinin binding to a histidine kinase receptor in the endoplasmic reticulum membrane.

This results in the autophosphorylation of the receptor, with the phosphate then being transferred to a phosphotransfer protein. The phosphotransfer proteins can then phosphorylate the type-B response regulators (RR) which are a family of transcriptions factors. The phosphorylated, and thus activated, type-B RRs regulate the transcription of numerous genes, including the type-A RRs. The type-A RRs negatively regulate the pathway.

Biosynthesis

Adenosine phosphate-isopentenyltransferase (IPT) catalyses the first reaction in the biosynthesis of isoprene cytokinins. It may use ATP, ADP, or AMP as substrates and may use dimethylallyl pyrophosphate (DMAPP) or hydroxymethylbutenyl pyrophosphate (HMBPP) as prenyl donors. This reaction is the rate-limiting step in cytokinin biosynthesis. DMADP and HMBDP used in cytokinin biosynthesis are produced by the methylerythritol phosphate pathway (MEP).

Cytokinins can also be produced by recycled tRNAs in plants and bacteria. tRNAs with anticodons that start with a uridine and carrying an already-prenylated adenosine adjacent to the anticodon release on degradation the adenosine as a cytokinin. The prenylation of these adenines is carried out by tRNA-isopentenyltransferase.

Auxin is known to regulate the biosynthesis of cytokinin.

Uses

Because cytokinin promotes plant cell division and growth, produce farmers use it to increase crops. One study found that applying cytokinin to cotton seedlings led to a 5–10% yield increase under drought conditions.

Cytokinins have recently been found to play a role in plant pathogenesis. For example, cytokinins have been described to induce resistance against Pseudomonas syringae in Arabidopsis thaliana and Nicotiana tabacum. Also in context of biological control of plant diseases cytokinins seem to have potential functions. Production of cytokinins by Pseudomonas fluorescens G20-18 has been identified as a key determinant to efficiently control the infection of A. thaliana with P. syringae.

Strigolactone

Strigolactones are a group of chemical compounds produced by a plant's roots. Due to their mechanism of action, these molecules have been classified as plant hormones or phytohormones. So far, strigolactones have been identified to be responsible for three different physiological processes: First, they promote the germination of parasitic organisms that grow in the host plant's roots, such as Striga lutea and other plants of the genus Striga. Second, strigolactones are fundamental for the recognition of the plant by symbiotic fungi, especially arbuscular mycorrhizal fungi, because they establish a mutualistic association with these plants, and provide phosphate and other soil nutrients. Third, strigolactones have been identified as branching inhibition hormones in plants; when present, these compounds prevent excess bud growing in stem terminals, stopping the branching mechanism in plants.

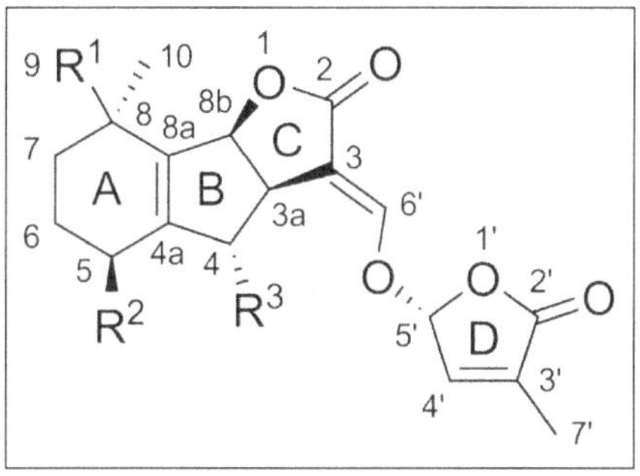

General structure of strigolactones.

Strigolactones comprise a diverse group, but they all have core common chemical structure, as shown in the image to the right. The structure is based on a tricyclic lactone linked to a hydroxymethyl butenolide; the former is represented in the figure as the A-B-C part, while the latter is the D part of the molecule. It is important to note that most strigolactones present variations in the ABC part, but the D ring is quite constant across the different species, which led researchers to suspect that the biological activity relies on this part of the molecule. Different studies have demonstrated that the activity of the molecules is lost when the C-D section of the molecules is modified.

Since strigolactones are involved in the signaling pathway required for germination of parasitic species (such as Striga sp.), they have been a proposed target to control pests and overgrowth of these parasitic organism. Using a molecule similar to strigolactones could be the key to designing a chemical and biological mechanism to stop the colonization of avplant's root by parasitic plants.

Chemistry

Properties

Although strigolactones vary in some of their functional groups, their melting point is usually found always between 200 and 202 Celsius degrees. The decomposition of the molecule occurs after reaching 195°C. They are highly soluble in polar solvents, such as acetone; soluble in benzene, and almost insoluble in hexane.

Chemical Structures

Some examples of strigolactones include:

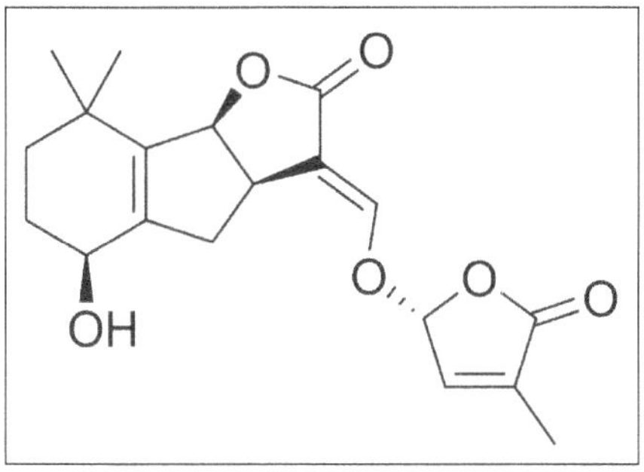

(+)-Strigol.

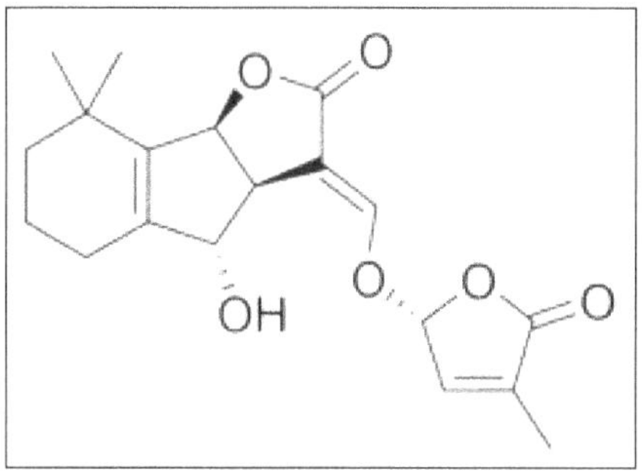

(+)-Orobanchol.

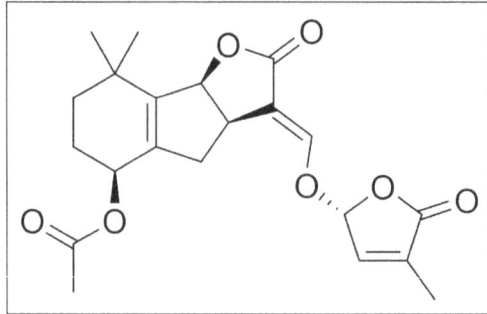

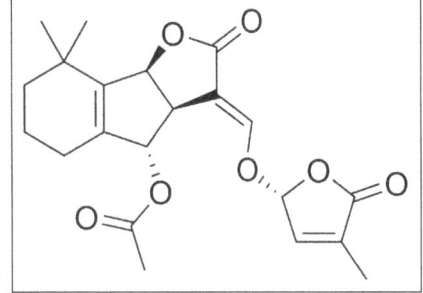

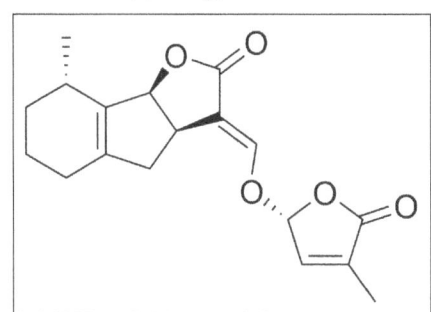

(+)-5-Deoxystrigol.   (+)-Strigyl acetate.

(+)-Orobanchyl acetate.   Sorgolactone.

Biosynthesis

Carotenoid Pathway via Carlactone

The biosynthetic pathway of the strigolactones has not been fully elucidated, but different steps have been identified, including the required enzymes to carry out the chemical transformation. The first step is the isomerization of the 9th chemical bond of the β-carotene, changing from trans configuration to cis. This first step is carried out by the enzyme β-carotene isomerase, also called DWARF27 or D27 for short, which required iron as a cofactor. The second step is the chemical separation of 9-cis-β-carotene into two different compounds: the first one is 9-cis-aldehyde and the second is β-ionone. This second step is catalized by the carotenoid cleavage deoxygenase 7 (CCD7). In the third step, another carotenoid cleavage oxygenase, called CCD8 (from the same family as CCD7), catalyze the conversion and rearrangement of the aldehyde created in the previous step into 9-cis-β-apo-10 and subsequently producing carlactone.

It is still not clear how exactly carlactone is transformed into the different strigolactones identified so far, but several studies have proved that carlactone is definitely the precursor of strigolactones. This last step of the biosynthesis should involve the addition of at least two oxygen molecules to convert the carlactone in 5-deoxystrigol, a simple strigolactone, and more oxidation should be required to produce other more complex strigolactone. The protein MAX1 has been proposed to catalyze the last step of the biosynthesis of strigolactones due its role in oxidative metabolism in plants.

Role of ABA in Biosynthesis

Both, abscisic acid (ABA) and strigolactones have a common group of enzymes that carried out the synthesis of the two compounds, previously it had been demonstrated the existence of a correlation of the two biosynthesis pathways, and it has been supported by different studies. The ABA biosynthesis relies in a set of enzymes, called 9-cis-epoxycarotenoid dyoxygenase (NCED). But, mutants plants that were defective in the production of the NCED enzymes, not just presented low levels of ABA, rather they also present low levels of strigolactones, specifically in the roots extracts where this hormone is mostly synthesized, this finding provided the basis for the existence of a common enzymatic machinery, Other experiments that consist in blocking the NCED enzymes and using mutants unable to detect ABA changes, were used to support this theory. So far there is a clear correlation of both synthesis that is related to the used of NCED enzymes in both biosynthesis, but the exact mechanism in which they are connected remains unclear.

Molecular Perception

In plants, strigolactones are perceived by the dual receptor/hydrolase protein DWARF14 (D14), a member of the α/β hydrolase superfamily. Despite being considered hydrolases with poor substrate turnover, an intact catalytic triad is required for the protein's biological function. Several (in part competing) models have been proposed for the involvement of the catalytic triad in ligand perception:

- Hydrolysis of strigolactone, resulting in the D-ring being covalently attached to the active site serine.

- Hydrolysis of strigolactone, resulting in a free D-ring that serves as a molecular glue at the entrance of the receptor, mediating interaction with another protein.

- Binding of non-hydrolyzed, intact strigolactone that generates an altered DWARF14 protein surface, mediating interaction with another protein.

- Hydrolysis of strigolactone, resulting in the D-ring being covalently attached to the active site histidine.

- Hydrolysis of strigolactone, resulting in the D-ring being covalently attached to the active site serine and histidine at the same time, inducing a conformational change of DWARF14, leading to interaction with another protein.

Kinetic results have suggested that the intact strigolactone triggers a signaling cascade after which hydrolysis is carried out as the final step to inactivate the strigolactone molecule.

Mechanism of Action

Germination of Arbuscular Mycorrhiza

Strigolactones are known to stimulate the germination of arbuscular mycorrhiza spores. Since they produce this effect at extremely low concentrations, it has been proposed that the mechanism of activation must be a signaling pathway. Different studies with diverse type of fungi, have found that after stimulation with strigolactones, the fungal cells present a higher amount of mitochondria and an increase in their oxidative activity. Due the role of mitochondria in oxidative metabolism of macronutrients, it is thought that the spores remain inactive before finding the host plant, and once they are stimulated with strigolactones, the oxidative machinery in the mitochondrion gets activated to produce energy and nutrients necessaries for germination of the spore and fungal branching. Studies with root extracts support this hypothesis, and so far strigolactones are the candidate molecules that better explain this increased in mitochondrial activity.

Auxin-mediated Secondary Growth

It has been established that secondary growth in plant is mainly regulated by the phytohormone auxin. However, the mechanism of auxin secretion is at the same time regulated by strigolactones, thus the latter can control secondary growth through auxin. When applied in terminal buds of stem, strigolactone can block the expression of transport proteins required to move auxin across the buds, these proteins are denominated PIN1. Thus, it was not surprising that when analyzing strigolactone deficient mutants, they were found to present an over-expression of PIN1 protein, which facilitate the transport of auxin in the terminal buds; auxin prevented the mitotic activity of these buds, stopping the plant to initiate secondary growth and branching. In conclusion, plants depend in auxin transport for secondary growth initiation or inhibition, but these transport mechanism is dependent of the production of strigolactones, which can easily travel from the site of production (roots) to the terminal buds of the stem through the xylem.

Abscisic Acid

Abscisic acid (ABA) is a plant hormone. ABA functions in many plant developmental processes, including seed and bud dormancy, the control of organ size and stomatal closure. It is especially important for plants in the response to environmental stresses, including drought, soil salinity, cold tolerance, freezing tolerance, heat stress and heavy metal ion tolerance.

In Plants

Function

ABA was originally believed to be involved in abscission. This is now known to be the case only in a small number of plants. ABA-mediated signaling also plays an important part in plant responses to environmental stress and plant pathogens. The plant genes for ABA biosynthesis and sequence of the pathway have been elucidated. ABA is also produced by some plant pathogenic fungi via a biosynthetic route different from ABA biosynthesis in plants.

Abscisic acid owes its names to its role in the abscission of plant leaves. In preparation for winter, ABA is produced in terminal buds. This slows plant growth and directs leaf primordia to develop scales to protect the dormant buds during the cold season. ABA also inhibits the division of cells in the vascular cambium, adjusting to cold conditions in the winter by suspending primary and secondary growth.

Abscisic acid is also produced in the roots in response to decreased soil water potential (which is associated with dry soil) and other situations in which the plant may be under stress. ABA then translocates to the leaves, where it rapidly alters the osmotic potential of stomatal guard cells, causing them to shrink and stomata to close. The ABA-induced stomatal closure reduces transpiration (evaporation of water out of the stomata), thus preventing further water loss from the leaves in times of low water availability. A close linear correlation was found between the ABA content of the leaves and their conductance (stomatal resistance) on a leaf area basis.

Seed germination is inhibited by ABA in antagonism with gibberellin. ABA also prevents loss of seed dormancy.

Several ABA-mutant Arabidopsis thaliana plants have been identified and are available from the Nottingham Arabidopsis Stock Centre - both those deficient in ABA production and those with altered sensitivity to its action. Plants that are hypersensitive or insensitive to ABA show phenotypes in seed dormancy, germination, stomatal regulation, and some mutants show stunted growth and brown/yellow leaves. These mutants reflect the importance of ABA in seed germination and early embryo development.

Pyrabactin (a pyridyl containing ABA activator) is a naphthalene sulfonamide hypocotyl cell expansion inhibitor, which is an agonist of the seed ABA signaling pathway. It is the first agonist of the ABA pathway that is not structurally related to ABA.

Homeostasis

Biosynthesis

Abscisic acid (ABA) is an isoprenoid plant hormone, which is synthesized in the plastidal 2-C-methyl-D-erythritol-4-phosphate (MEP) pathway; unlike the structurally related sesquiterpenes, which are formed from the mevalonic acid-derived precursor farnesyl diphosphate (FDP), the C_{15} backbone of ABA is formed after cleavage of C_{40} carotenoids in MEP. Zeaxanthin is the first committed ABA precursor; a series of enzyme-catalyzed epoxidations and isomerizations via violaxanthin, and final cleavage of the C_{40} carotenoid by a dioxygenation reaction yields the proximal ABA precursor, xanthoxin, which is then further oxidized to ABA. via abscisic aldehyde.

Abamine has been designed, synthesized, developed and then patented as the first specific ABA biosynthesis inhibitor, which makes it possible to regulate endogenous levels of ABA.

Location and Timing of ABA Biosynthesis

- Released during desiccation of the vegetative tissues and when roots encounter soil compaction.
- Synthesized in green fruits at the beginning of the winter period.
- Synthesized in maturing seeds, establishing dormancy.
- Mobile within the leaf and can be rapidly translocated from the roots to the leaves by the transpiration stream in the xylem.

- Produced in response to environmental stress, such as heat stress, water stress, salt stress.

- Synthesized in all plant parts, e.g., roots, flowers, leaves and stems.

- ABA is synthesised in almost all cells that contain chloroplasts or amyloplasts.

Inactivation

ABA can be catabolized to phaseic acid via CYP707A (a group of P450 enzymes) or inactivated by glucose conjugation (ABA-glucose ester) via the enzyme AOG. Catabolism via the CYP707As is very important for ABA homeostasis, and mutants in those genes generally accumulate higher levels of ABA than lines overexpressing ABA biosynthetic genes. In soil bacteria, an alternative catabolic pathway leading to dehydrovomifoliol via the enzyme vomifoliol dehydrogenase has been reported.

Effects

- Antitranspirant: In drought prone areas, water stress is serious problem in agriculture production. so sprays of ABA are suggested that cause partial closure of stomata for few days, to reduce transpirational loss of water.

- Inhibits fruit ripening.

- Inhibits seed germination: Responsible for seed dormancy by inhibiting cell growth.

- Inhibits the synthesis of Kinetin nucleotide.

- Downregulates enzymes needed for photosynthesis.

- Acts on endodermis to prevent growth of roots when exposed to salty conditions.

- Delays cell division.

- Dormancy inducer: It is used to induce dormancy in the seeds.

Signal Cascade

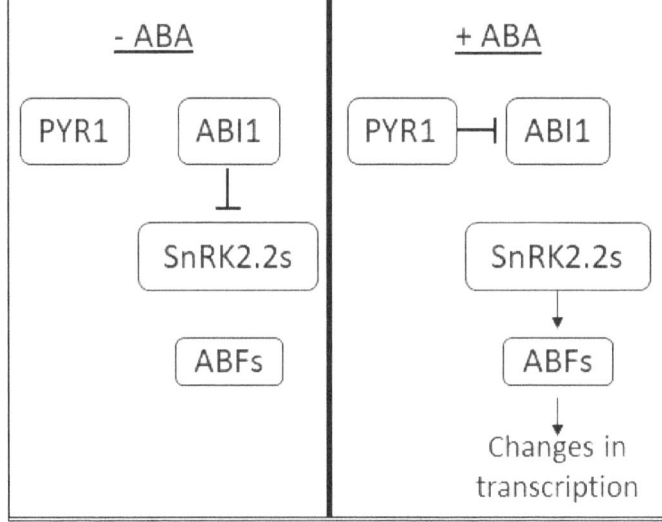

ABA signal pathway in plants.

In the absence of ABA, the phosphatase ABI1-INSENSITIVE1 (ABI1) inhibits the action of SNF1-related protein kinases (subfamily 2) (SnRK2s). ABA is perceived by the PYRABACTIN RESISTANCE 1 (PYR1) and PYR1-like membrane proteins. On ABA binding, PYR1 binds to and inhibits ABI1. When SnRK2s are released from inhibition, they activate several transcription factors from the ABA RESPONSIVE ELEMENT-BINDING FACTOR (ABF) family. ABFs then go on to cause changes in the expression of a large number of genes. Around 10% of plant genes are thought to be regulated by ABA.

Plant Peptide Hormone

Peptide signaling plays a significant role in various aspects of plant growth and development and specific receptors for various peptides have been identified as being membrane-localized receptor kinases, the largest family of receptor-like molecules in plants. Signaling peptides include members of the following protein families.

- Systemin — is a small polypeptide functioning as a long-distance signal to activate chemical defenses against herbivores. It was the first plant hormone proven to be a peptide. Systemin induces the production of protein defense compound called protease inhibitors. Systemin was first identified in tomato leaves. It was found to be an 18-amino acid peptide processed from the C-terminus of a 200-amino acid precursor, which is called prosystemin.

- CLV3/ESR-related ('CLE') peptide family — CLV3 encodes a small secreted peptide that functions as a short range ligand to the membrane-bound CLV1 receptor like kinase that together with CLV2 (a receptor-like protein) function to maintain stem cell homeostasis in Arabidopsis shoot apical meristems. Although the maize embryo-surrounding region protein (ESR). and CLV3 are very different, they are both members of the CLE peptide family given that they share a short conserved 14-amino acid sequence at the carboxy terminal region. To date, more than 150 CLE signaling peptides are identified. This proteolytically processed bioactive region is important for both promoting and inhibiting cellular differentiation in both apical and cambial meristems. Recently it was found that CLE25 can act as a long distance signal to communicate water stress from the roots to leaves.

- ENOD40 — is an early nodulin gene, hence ENOD, that putatively encodes two small peptides, one of 12 and the other of 18 amino acid residues. Controversy exists on whether the mRNA or peptides themselves are responsible for bioactivity. Both peptides have been shown "in vivo" to bind to the 93 kDa subunit of sucrose synthase, an essential component in sucrose metabolism. Sucrose degradation is a key step in nitrogen fixation, and is a pre-requisite for normal nodule development.

- Phytosulfokine (PSK) — was first identified as a "conditioning factor" in asparagus and carrot cell cultures. The bioactive five amino acid peptide (PSK) is proteolytically processed from an ~80 amino acid precursor secreted peptide. PSK has been demonstrated to promote cellular proliferation and transdifferentiation. It has been demonstrated that PSK binds to a membrane bound LRR receptor like kinase (PSKR).

- POLARIS (PLS) — The PLS peptide has a predicted length of 36 amino acids however possesses no secretion signal, suggesting that it functions within the cytoplasm. The PLS peptide itself has not yet been biochemically isolated, however loss-of-function mutants are hypersensitive to cytokinin with reduced responsiveness to auxin. Developmentally it is involved in vascularization, longitudinal cell expansion and increased radial expansion.

- Rapid Alkalinization Factor (RALF) — is 49 amino acid peptide that was identified whilst purifying systemin from tobacco leaves, it causes rapid medium alkanalization and does not activate defence responses like systemin. Tomato RALF precursor cDNA encodes a 115 amino acid polypeptide containing an amino-terminal signal sequence with the bioactive RALF peptide encoded at the carboxy terminus. It is not known how mature RALF peptide is produced from its precursor, but a dibasic amino acid motif (typical of recognition sites of processing enzymes in yeast and animals) is located two residues upstream from the amino terminus of mature RALF. RALF has been identified to bind to potential membrane bound receptors complex containing proteins 25 kDa and 120 kDa in size.

- SCR/SP11 — are small polymorphic peptides produced by the tapetal cells of anthers and is involved in self-incompatibility of Brassica species. This secreted polypeptide is between 78 and 80 amino acid residues in length. Unlike other peptide hormones, no further post-translational processing occurs, except for the removal of the N-terminal signal peptide. SCR/SP11 like other small peptide hormones binds to a membrane bound LRR receptor like kinase (SRK).

- ROTUNDIFOLIA4/DEVIL1 (ROT4/DVL1) — The ROT4 and DVL1 are peptides of 53 and 51 amino acids respectively, which have a high degree of sequence homology. They are two members of 23 member peptide family. ROT4 and DVL1 are involved in regulating polar cell proliferation on the longitudinal axis of organs.

- Inflorescence deficient in abscission (IDA) — a family of secreted peptides identified to be involved in petal abscission. The peptides are 77 amino acids in length and possess an amino-terminal secretions signal. Like the CLE peptide family these proteins have a conserved carboxy-terminal domain that is bordered by potentially cleavable basic residues. These proteins are secreted from cells in the floral abscission zone. Studies suggest that the HAESA membrane-associated LRR-RLK is likely to be this peptide's receptor as it too is expressed in the zone of floral organ abscission.

Jasmonate

Jasmonate (JA) and its derivatives are lipid-based plant hormones that regulate a wide range of processes in plants, ranging from growth and photosynthesis to reproductive development. In particular, JAs are critical for plant defense against herbivory and plant responses to poor environmental conditions and other kinds of abiotic and biotic challenges. Some JAs can also be released as volatile organic compounds (VOCs) to permit communication between plants in anticipation of mutual dangers.

The isolation of methyl jasmonate from jasmine oil derived from Jasminum grandiflorum led to the discovery of the molecular structure of jasmonates and their name.

Jasminum grandiflorum.

Chemical Structure

Structures of Active Jasmonate Derivatives

Jasmonates (JA) are an oxylipin, i.e. a derivative of oxygenated fatty acid. It is biosynthesized from linolenic acid in chloroplast membranes. Synthesis is initiated with the conversion of linolenic acid to 12-oxo-phytodienoic acid (OPDA), which then undergoes a reduction and three rounds of oxidation to form (+)-7-iso-JA, jasmonic acid. Only the conversion of linolenic acid to OPDA occurs in the chloroplast; all subsequent reactions occur in the peroxisome.

Jasmonic acid (JA).   Methyl JA.

JA itself can be further metabolized into active or inactive derivatives. Methyl JA (MeJA) is a volatile compound that is potentially responsible for interplant communication. JA conjugated with amino acid isoleucine (Ile) results in JA-Ile, which is currently the only known JA derivative needed for JA signaling. JA undergoes decarboxylation to give cis-jasmone.

Mechanism of Signaling

In general, the steps in jasmonate (JA) signaling mirror that of auxin signaling: the first step comprises E3 ubiquitin ligase complexes, which tag substrates with ubiquitin to mark them for degradation by proteasomes. The second step utilizes transcription factors to effect physiological changes. One of the key molecules in this pathway is JAZ, which serves as the on-off switch for JA signaling. In the absence of JA, JAZ proteins bind to downstream transcription factors and limit their activity. However, in the presence of JA or its bioactive derivatives, JAZ proteins are degraded, freeing transcription factors for expression of genes needed in stress responses.

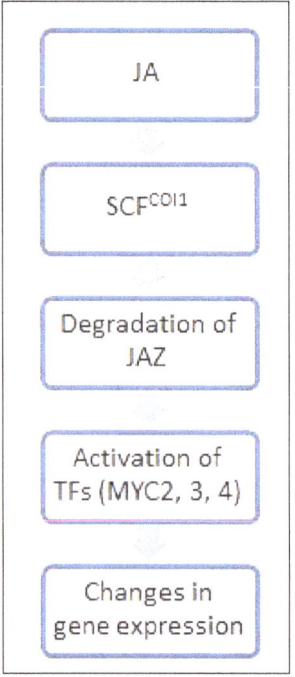

Major components of the jasmonate pathway.

Because JAZ did not disappear in null coi1 mutant plant backgrounds, protein COI1 was shown to mediate JAZ degradation. COI1 belongs to the family of highly conserved F-box proteins, and it recruits substrates for the E3 ubiquitin ligase SCF^{COI1}. The complexes that ultimately form are known as SCF complexes. These complexes bind JAZ and target it for proteasomal degradation. However, given the large spectrum of JA molecules, not all JA derivatives activate this pathway for signaling, and the range of those participating in this pathway is unknown. Thus far, only JA-Ile has been shown to be necessary for COI1-mediated degradation of JAZ11. JA-Ile and structurally related derivatives can bind to COI1-JAZ complexes and promote ubiquitination and thus degradation of the latter.

This mechanistic model raises the possibility that COI1 serves as an intracellular receptor for JA signals. Recent research has confirmed this hypothesis by demonstrating that the COI1-JAZ complex acts as a co-receptor for JA perception. Specifically, JA-Ile binds both to a ligand-binding pocket in COI1 and to a 20 amino-acid stretch of the conserved Jas motif in JAZ. This JAZ residue acts as a plug for the pocket in COI1, keeping JA-Ile bound in the pocket. Additionally, co-purification and subsequent removal of inositol pentakisphosphate (InsP$_5$) from COI1 suggest InsP$_5$ is a necessary component of the co-receptor and plays a role in potentiating the co-receptor complex.

Once freed from JAZ, transcription factors can activate genes needed for a specific JA response. The best-studied transcription factors acting in this pathway belong to the MYC family of transcription factors, which are characterized by a basic helix-loop-helix (bHLH) DNA binding motif.

These factors (of which there are three, MYC2, 3, and 4) tend to act additively. For example, a plant that has only lost one myc becomes more susceptible to insect herbivory than a normal plant. A plant that has lost all three will be as susceptible to damage as coi1 mutants, which are completely unresponsive to JA and cannot mount a defense against herbivory. However, while all these MYC molecules share functions, they vary greatly in expression patterns and transcription functions. For instance, MYC2 has a greater effect on root growth compared to MYC3 or MYC4.

Additionally, MYC2 will loop back and regulate JAZ expression levels, leading to a negative feedback loop. These transcription factors all have different impacts on JAZ levels after JA signaling. JAZ levels in turn affect transcription factor and gene expression levels. In other words, on top of activating different response genes, the transcription factors can vary JAZ levels to achieve specificity in response to JA signals.

Function

Although jasmonate (JA) regulates many different processes in the plant, its role in wound response is best understood. Following mechanical wounding or herbivory, JA biosynthesis is rapidly activated, leading to expression of the appropriate response genes. For example, in the tomato, wounding produces defense molecules that inhibit leaf digestion in the insect's gut. Another indirect result of JA signaling is the volatile emission of JA-derived compounds. MeJA on leaves can travel airborne to nearby plants and elevate levels of transcripts related to wound response. In general, this emission can further upregulate JA synthesis and signaling and induce nearby plants to prime their defenses in case of herbivory.

Following its role in defense, JAs have also been implicated in cell death and leaf senescence. JA can interact with many kinases and transcription factors associated with senescence. JA can also induce mitochondrial death by inducing the accumulation of reactive oxygen species (ROSs). These compounds disrupt mitochondria membranes and compromise the cell by causing apoptosis, or programmed cell death. JAs' roles in these processes are suggestive of methods by which the plant defends itself against biotic challenges and limits the spread of infections.

JA and its derivatives have also been implicated in plant development, symbiosis, and a host of other processes included in the list below:

- By studying mutants overexpressing JA, one of the earliest discoveries made was that JA inhibits root growth. The mechanism behind this event is still not understood, but mutants in the COI1-dependent signaling pathway tend to show reduced inhibition, demonstrating that the COI1 pathway is somehow necessary for inhibiting root growth.

- JA plays many roles in flower development. Mutants in JA synthesis or in JA signaling in Arabidopsis present with male sterility, typically due to delayed development. The same genes promoting male fertility in Arabidopsis promote female fertility in tomatoes. Overexpression of 12-OH-JA can also delay flowering.

- JA and MeJA inhibit the germination of nondormant seeds and stimulate the germination of dormant seeds.

- High levels of JA encourage the accumulation of storage proteins; genes encoding vegetative storage proteins are JA responsive. Specifically, tuberonic acid, a JA derivative, induces the formation of tubers.

- JAs also play a role in symbiosis between plants and microorganisms; however, its precise role is still unclear. JA currently appears to regulate signal exchange and nodulation regulation between legumes and rhizobium. On the other hand, elevated JA levels appear to regulate carbohydrate partitioning and stress tolerance in mycorrhizal plants.

Role in Pathogenesis

Pseudomonas syringae causes bacterial speck disease in tomatoes by hijacking the plant's jasmonate (JA) signaling pathway. This bacteria utilizes a type III secretion system to inject a cocktail of viral effector proteins into host cells.

One of the molecules included in this mixture is the phytotoxin coronatine (COR). JA-insensitive plants are highly resistant to P. syringae and unresponsive to COR; additionally, applying MeJA was sufficient to rescue virulence in COR mutant bacteria. Infected plants also expressed downstream JA and wound response genes but repressed levels of pathogenesis-related (PR) genes. All these data suggest COR acts through the JA pathway to invade host plants. Activation of a wound response is hypothesized to come at the expense of pathogen defense. By activating the JA wound response pathway, P. syringae could divert resources from its host's immune system and infect more effectively.

Plants produce N-acylamides that confer resistance to necrotrophic pathogens by activating JA biosynthesis and signalling. Arachidonic acid (AA), the counterpart of the JA precursor α-LeA occurring in metazoan species but not in plants, is perceived by plants and acts through an increase in JA levels concomitantly with resistance to necrotrophic pathogens. AA is an evolutionarily conserved signalling molecule that acts in plants in response to stress similar to that in animal systems.

Cross Talk with other Defense Pathways

While the jasmonate (JA) pathway is critical for wound response, it is not the only signaling pathway mediating defense in plants. To build an optimal yet efficient defense, the different defense pathways must be capable of cross talk to fine-tune and specify responses to abiotic and biotic challenges.

One of the best studied examples of JA cross talk occurs with salicylic acid (SA). SA, a hormone, mediates defense against pathogens by inducing both the expression of pathogenesis-related genes and systemic acquired resistance (SAR), in which the whole plant gains resistance to a pathogen after localized exposure to it.

Wound and pathogen response appear to be interact negatively. For example, silencing phenylalanine ammonia lyase (PAL), an enzyme synthesizing precursors to SA, reduces SAR but enhances herbivory resistance against insects. Similarly, overexpression of PAL enhances SAR but reduces wound response after insect herbivory. Generally, it has been found that pathogens living in live plant cells are more sensitive to SA-induced defenses, while herbivorous insects and pathogens that derive benefit from cell death are more susceptible to JA defenses. Thus, this trade-off in pathways optimizes defense and saves plant resources.

Cross talk also occurs between JA and other plant hormone pathways, such as those of abscisic acid (ABA) and ethylene (ET). These interactions similarly optimize defense against pathogens and herbivores of different lifestyles. For example, MYC2 activity can be stimulated by both JA and ABA pathways, allowing it to integrate signals from both pathways. Other transcription factors such as ERF1 arise as a result of JA and ET signaling. All these molecules can act in combination to activate specific wound response genes.

Finally, cross talk is not restricted for defense: JA and ET interactions are critical in development as well, and a balance between the two compounds is necessary for proper apical hook development in Arabidopsis seedlings.

Karrikin

Karrikins are a group of plant growth regulators found in the smoke of burning plant material. For many years smoke from wildfires or bushfires was known to stimulate the germination of seeds. In 2004, butenolide was shown to be responsible for this effect. Later, several closely related compounds were discovered in smoke, and are collectively known as karrikins. Six karrikins have been discovered in smoke and are designated KAR_1, KAR_2, KAR_3, KAR_4, KAR_5 and KAR_6, but KAR_1 to KAR_4 are the most active. The butenolide part of the compound is a 5-membered lactone ring while the other part of the karrikin compound is a 6-membered pyran ring.

Chemical structures of the karrikins.

Karrikins are formed by the heating or combustion of carbohydrates including sugars and cellulose. When plant material burns, these carbohydrates convert to karrikins. Plant cell walls are made of polysaccharides including cellulose, and a convenient way to prepare karrikins is to burn paper or dried grass. The pyran part of karrikins is probably directly derived from a pyranose sugar. There is no evidence that karrikins occur naturally in plants, but it has been postulated that karrikin-like molecules do.

The Response to Karrikins

Karrikins produced by bushfires occur largely in the ash at the site of the fire. Rains occurring after the fire wash the karrikins into the soil where dormant seeds reside. The karrikins and water can provide a 'wake-up call' for such seeds, triggering germination of the soil seed bank. The plants that emerge grow quickly, flower and produce new seeds, which fall to the ground. These seeds can remain in the soil for decades, until the next fire produces fresh karrikins. Plants with this lifestyle are known as fire ephemerals. They thrive because the fire removes competing vegetation and provides nutrients and light for the emerging seedlings. Plants in many families respond to smoke and karrikins, suggesting that this response has evolved independently in different groups.

Perception of Karrikins

The mode of action of karrikins has been largely determined using the genetic resources of Arabidopsis thaliana. Perception of karrikins by Arabidopsis requires an alpha/beta-fold hydrolase named KARRIKIN-INSENSITIVE-2 (KAI2). The KAI2 protein has a catalytic triad of amino acids which is essential for activity, consistent with the hypothesis that KAI2 hydrolyses its ligand. This model is consistent with the perception of the chemically related strigolactone hormones which involves hydrolysis by their receptor protein DWARF14, an alpha/beta hydrolase related to KAI2. The question of whether karrikins act directly in plants is controversial. While some studies suggest that karrikins can bind directly to KAI2 protein, others do not support this. It is possible that karrikins produced by wildfires are converted to a different compound by the plant, before interaction with KAI2. The ability of different plants to carry out this conversion could partly explain differences in their ability to respond to karrikins and to smoke.

Karrikin Signalling

The activity of karrikins requires an F-box protein named MORE AXILLARY GROWTH-2 (MAX2) in Arabidopsis. This protein is also required for strigolactone signaling in Arabidopsis. Homologs of MAX2 are also required for strigolactone signaling in rice (known as DWARF3) petunia (DAD2) and pea (RMS4). Karrikin signaling also requires a protein named SUPPRESSOR OF MORE AXILLARY GROWTH2-1 (SMAX1) which is a homolog of the DWARF53 protein required for strigolactone signaling in rice. SMAX1 and DWARF53 proteins could be involved in the control of cellular functions such as transport or transcription. The present model for karrikin and strigolactone signaling involves interaction of KAI2 or DWARF14 with SMAX1 or DWARF53 proteins respectively, which targets those proteins for ubiquitination and destruction.

Effects of Karrikins on Plant Growth

Karrikins not only stimulate seed germination, but are reported to increase seedling vigour. In Arabidopsis, karrikins influence seedling photomorphogenesis, resulting in shorter hypocotyls and larger cotyledons. Such responses could provide seedlings with an advantage as they emerge into the post-fire landscape. The KAI2 protein is also required for leaf development, implying that karrikins could influence other aspects of plant growth.

Evolution of Karrikin Response

The gene for KAI2 protein is present in lower plants including algae and mosses, whereas the DWARF14 protein evolved with seed plants, probably as a result of duplication of KAI2 followed by functional specialisation. Karrikin signaling could have evolved with seed plants as a result of the divergence of KAI2 and DWARF14 functions, possibly during the Cretaceous period when fires were common on Earth.

The Endogenous Signal for KAI2

Karrikins are produced by wildfires but all seed plants contain KAI2 proteins, raising the question of the usual function of this protein. There is compelling evidence that plants contain an endogenous compound that is perceived by KAI2 to control seed germination and plant development, but this compound is neither a karrikin nor a strigolactone.

Florigen

Florigen (or flowering hormone) is the hypothesized hormone-like molecule responsible for controlling and/or triggering flowering in plants. Florigen is produced in the leaves, and acts in the shoot apical meristem of buds and growing tips. It is known to be graft-transmissible, and even functions between species. However, despite having been sought since the 1930s, the exact nature of florigen is still a mystery.

Mechanism

Central to the hunt for florigen is an understanding of how plants use seasonal changes in day length to mediate flowering—a mechanism known as photoperiodism. Plants which exhibit photoperiodism may be either 'short day' or 'long day' plants, which in order to flower require short days or long days respectively, although plants in fact distinguish day length from night length.

The current model suggests the involvement of multiple different factors. Research into florigen is predominately centred on the model organism and long day plant, Arabidopsis thaliana. Whilst much of the florigen pathways appear to be well conserved in other studied species, variations do exist. The mechanism may be broken down into three stages: photoperiod-regulated initiation, signal translocation via the phloem, and induction of flowering at the shoot apical meristem.

Initiation

In Arabidopsis thaliana, the signal is initiated by the production of messenger RNA (mRNA) coding a transcription factor called CONSTANS (CO). CO mRNA is produced approximately 12 hours after dawn, a cycle regulated by the plant's biological clock. This mRNA is then translated into CO protein. However CO protein is stable only in light, so levels stay low throughout short days and are only able to peak at dusk during long days when there is still a little light. CO protein promotes transcription of another gene called Flowering Locus T (FT). By this mechanism, CO protein may only reach levels capable of promoting FT transcription when exposed to long days. Hence, the transmission of florigen—and thus, the induction of flowering—relies on a comparison between the plant's perception of day/night and its own internal biological clock.

Translocation

The FT protein resulting from the short period of CO transcription factor activity is then transported via the phloem to the shoot apical meristem.

Flowering

At the shoot apical meristem, the FT protein interacts with a transcription factor (FD protein) to activate floral identity genes, thus inducing flowering. Specifically, arrival of FT at the shoot apical meristem and formation of the FT/FD heterodimer is followed by the increased expression of at least one direct target gene, APETALA 1 (AP1), along with other targets, such as SOC1 and several SPL genes, which are targeted by a microRNA.

Research

Florigen was first described by Soviet Armenian plant physiologist Mikhail Chailakhyan, who in 1937 demonstrated that floral induction can be transmitted through a graft from an induced plant to one that has not been induced to flower. Anton Lang showed that several long-day plants and biennials could be made to flower by treatment with gibberellin, when grown under a non-flower-inducing (or non-inducing) photoperiod. This led to the suggestion that florigen may be made up of two classes of flowering hormones: Gibberellins and Anthesins. It was later postulated that during non-inducing photoperiods, long-day plants produce anthesin, but no gibberellin while short-day plants produce gibberellin but no anthesin. However, these findings did not account for the fact that short-day plants grown under non-inducing conditions (thus producing gibberellin) will not cause flowering of grafted long-day plants that are also under noninductive conditions (thus producing anthesin).

As a result of the problems with isolating florigen, and of the inconsistent results acquired, it has been suggested that florigen does not exist as an individual substance; rather, florigen's effect could be the result of a particular ratio of other hormones. However, more recent findings indicate that florigen does exist and is produced, or at least activated, in the leaves of the plant and that this signal is then transported via the phloem to the growing tip at the shoot apical meristem where the signal acts by inducing flowering. In Arabidopsis thaliana, some researchers have identified this signal as mRNA coded by the FLOWERING LOCUS T (FT) gene, others as the resulting FT protein. However, in 2007 other group of scientists made a breakthrough saying that it is not the mRNA, but the FT Protein that is transmitted from leaves to shoot possibly acting as "Florigen". The initial article that described FT mRNA as flowering stimuli was retracted by the authors themselves.

Role of GI, CO, FT Genes and Ca $^{2+}$/CaM

There are three genes involved in clock-controlled flowering pathway, GIGANTEA (GI), CONSTANS (CO), and FLOWERING LOCUS T (FT). Constant overexpression of GI from the Cauliflower mosaic virus 35S promoter causes early flowering under short day so an increase in GI mRNA expression induces flowering. Also, GI increases the expression of FT and CO mRNA, and FT and CO mutants showed later flowering time than GI mutant. In other words, functional FT and CO genes are required for flowering under short day. In addition, these flowering genes accumulate during light phase and decline during dark phase, which are measured by green fluorescent protein. Thus, their expressions oscillate during the 24-hour light-dark-cycle. In conclusion, the accumulation of GI mRNA alone or GI, FT, and CO mRNA promote flowering in Arabidopsis thaliana and these genes expressed in the temporal sequence GI-CO-FT.

Action potential triggers calcium flux into neurons in animal or root apex cells in plants. The intra-cellular calcium signals are responsible for regulation of many biological functions in organisms. For instance, Ca2+ binding to calmodulin, a Ca2+-binding proteins in animals and plants, controls gene transcriptions.

A Possible Mechanism for Flowering

A biological mechanism is proposed based on the information we have above. Light is the flowering signal of Arabidopsis thaliana. Light activates photo-receptors and triggers signal cascades in plant cells of apical or lateral meristems. Action potential is spread via the phloem to the root and more voltage-gated calcium channels are opened along the stem. There is increase in calcium ions influx in plant cells. These ions bind to calmodulin and the Ca2+/CaM signaling system triggers the expression of GI mRNA or FT and CO mRNA. The accumulation of GI mRNA or GI-CO-FT mRNA during the day causing the plant to flower.

Ethylene

Ethylene is a simple gaseous hydrocarbon produced from an amino acid and appears in most plant tissues in large amounts when they are stressed. It diffuses from its site of origin into the air and affects surrounding plants as well. Large amounts ordinarily are produced by roots, senescing flowers, ripening fruits, and the apical meristem of shoots. Auxin increases ethylene production, as does ethylene itself—small amounts of ethylene initiate copious production of still more. Ethylene stimulates the ripening of fruit and initiates abscission of fruits and leaves. In monoecious plants (those with separate male and female flowers borne on the same plant), gibberellins and ethylene concentrations determine the sex of the flowers: Flower buds exposed to high concentrations of ethylene produce carpellate flowers, while gibberellins induce staminate ones.

Functions of Ethylene

Growth

Ethylene inhibits longitudinal growth but stimulates transverse or horizontal growth and swelling of axis.

Gravity

It decreases the sensitivity to gravity. Roots become Apo-geotropic while stems turn positively geotropic. Leaves and flowers undergo drooping. The phenomenon is called epinasty. Seedlings develop tight epicotyl hook.

Senescence

It hastens the senescence of leaves and flowers.

Abscission

Abscission of various parts (leaves, flowers, fruits) is stimulated by ethylene which induces the formation of hydrolases.

Apical Dominance

Ethylene promotes apical dominance and prolongs dormancy of lateral buds.

Breaking of Dormancy

It breaks the dormancy of buds, seeds and storage organs.

Abscisic Acid

It seems that formation of abscisic acid in the leaves under conditions of water stress is mediated through ethylene.

Growth of Rice Seedling

Ethylene promotes rapid elongation of leaf bases and internodes in deep water rice plants. As a result leaves remain above water.

Root Initiation

In low concentration ethylene helps in root initiation, growth of lateral roots and root hairs. This increases the absorption surface of the plant roots.

Fruit Ripening

It aids in ripening of climacteric fruits and dehiscence of dry fruits. Climacteric fruits are fleshy fruits which show a sudden sharp rise of respiration rate at the time of ripening (respiratory climacteric). They are usually transported in green or unripe stage. Ethylene is used to induce artificial ripening of these fruits, e.g., Apple, Mango, Banana, etc.

Flowering

It stimulates flowering in Pineapple and related plants as well as mango though in other cases the gaseous hormone causes fading of flowers. This helps in synchronizing fruit set.

Sex Expression

Like auxins and cytokinins, ethylene has a feminizing effect on sex expression. The genetically male plants of Cannabis can be induced to produce female flowers in the presence of ethylene. The number of female flowers and hence fruit is enhanced in monoecious plants like Cucumber.

Uses of Ethylene

Ethylene regulates a number of physiological processes. Therefore, it is widely used PGR in agriculture. The common compound used for obtaining ethylene is ethophen or ethrel which is 2-chloroethyl phosphonic acid. In aqueous solution, ethophen is readily absorbed and transported to various parts. It releases ethylene slowly.

The Various Commercial Uses of Ethylene are as Follows:

Fruit Ripening

Kerosene lamps and hay were previously used for stimulating colour development and ripening of some fleshy fruits, e.g., Banana, Mango, Apple, and Tomato. The effect is due to ethylene. Ethylene lamps are now specifically used for this purpose.

Feminising Effect

External supply of very small quantity of ethylene increases the number of female flowers and hence fruits in Cucumber.

Sprouting of Storage Organs

Rhizomes, corms, tubers, seeds (e.g., Peanut) and other storage organs can be made to sprout early by exposing them to ethylene.

Thinning

Excess flowers and young fruits are thinned with the help of ethylene, e.g., Cotton, Cherry, and Walnut. It allows better growth of remaining fruits.

References

- Plant-hormones: biologydictionary.net, Retrieved 19 May, 2019
- Types-of-plant-hormones, plant-biology-growth-of-plants: cliffsnotes.com, Retrieved 18 August, 2019
- Ethylene-history-function-and-uses, plant-hormones- 44735: biologydiscussion.com, Retrieved 16 March, 2019
- Hedden P, Sponsel V (2015). "A Century of Gibberellin Research". Journal of Plant Growth Regulation. 34 (4): 740–60. doi:10.1007/s00344-015-9546-1. PMC 4622167. PMID 26523085.

5
Photosynthesis

Photosynthesis is the process by which plants prepare their food in the presence of sunlight, carbon dioxide and chlorophyll. It occurs as two types of reactions – light dependent reactions and light independent reactions. This chapter discusses about these reactions and related aspects of photosynthesis in detail.

Photosynthesis is the conversion of the energy of sunlight into chemical energy by living organisms. In most cases, the raw materials are carbon dioxide and water; the energy source is sunlight; and the end-products are oxygen and (energy rich) carbohydrates, for example sucrose and starch. However, there are some classes of bacteria that utilize a form of photosynthesis that does not produce oxygen (anoxygenic photosynthesis). Photosynthesis is arguably the most important biochemical pathway, since nearly all life depends on it. It is a complex process occurring in higher plants, phytoplankton, algae, and even such bacteria as the cyanobacteria.

The leaf is the primary site of photosynthesis in plants.

Photosynthetic organisms are also referred to as photoautotrophs, because they synthesize food directly from inorganic compounds using light energy. In green plants and algae, photosynthesis takes place in specialized cellular compartments called chloroplasts. In photosynthetic bacteria, which lack membrane-bound compartments, the reactions take place directly in the cell.

The essential function of photosynthesis in the biosphere attests to the interdependence of life. Although oxygen is, strictly defined, a waste product of photosynthesis reactions, the majority of organisms, including plants, utilize oxygen for cellular respiration. Moreover, heterotrophs, which include animals, fungi, and most bacteria, are unable to synthesize organic compounds from inorganic sources, and must rely on the (direct or indirect) consumption of plants and other autotrophs to obtain the organic substrates necessary for growth and development.

The ancestors of many current species are thought to have evolved in response to the oxygen catastrophe, a massive environmental change believed to have occurred about 2.4 billion years ago. At about that time apparently, evolving life forms developed photosynthetic capabilities and began producing molecular oxygen in such large quantities that it eventually caused an ecological crisis because oxygen was toxic to anaerobic organisms, the dominant life form of that period. In addition to being a crisis for anaerobic organisms, the period of the oxygen level explosion opened tremendous opportunity for those forms of life that could exploit the newly abundant gas as a potent source for metabolic energy.

Life had remained energetically limited until the widespread availability of oxygen. This breakthrough in metabolic evolution greatly increased the free energy supply to living organisms: today, more than 10^{17} kcal of free energy is stored annually by photosynthesis on earth, which corresponds to the fixation of more than 10^{10} tons of carbon into carbohydrates and other organic compounds.

In chemical terms, photosynthesis is an example of an oxidation-reduction process. In plants, photosynthesis uses light energy to power the oxidation of water (i.e., the removal of electrons), to produce molecular oxygen, hydrogen ions, and electrons. Most of the hydrogen ions and electrons are then transferred to carbon dioxide, which is reduced (i.e., it gains electrons) to organic products.

Specifically, carbon dioxide is reduced to make triose phosphate (G3P), which is generally considered the prime end-product of photosynthesis. It can be used as an immediate food nutrient, or combined and rearranged to form monosaccharide sugars, such as glucose, which can be transported to other cells or packaged for storage as an insoluble polysaccharide such as starch.

The general chemical equation for photosynthesis is often presented in simplified form as:

$$CO_{2(gas)} + 2H_2O_{(liquid)} + photons \rightarrow CH_2O_{(aqueous)} + H_2O + O_{2(gas)}$$

where (CH_2O) refers to the general formula for a carbohydrate.

However, a more general formula, that includes forms of photosynthesis that do not result in oxygen, is:

$$CO_{2(gas)} + 2H_2A + photons \rightarrow CH_2O + H_2O + 2A$$

with H_2A acting as the electron donor. It may be water or it may be something such as H_2S, as in the case of purple sulfur bacteria that yield sulfur as a product rather than oxygen.

Note, the source of the oxygen comes from water, not from the carbon dioxide.

The Site of Photosynthesis

Photosynthesis Occurs in the Chloroplasts of Green Plants and Algae

The reactions of photosynthesis occur in cellular subcompartments called chloroplasts, which themselves are further compartmentalized by inner and outer membranes separated by an inter-membrane space. The inner membrane's interior space, called the stroma, is filled with a fluid whose rich supply of enzymes supports light-dependent reactions of photosynthesis occurring inside stacks of membranous flattened sacs (thylakoids). The thylakoid stacks are called grana (singular: granum).

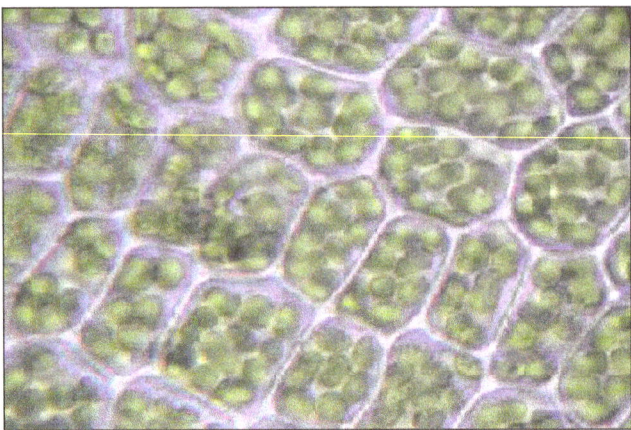

Plant cells with visible chloroplasts.

Embedded in the thylakoid membrane is the antenna complex comprising proteins and light-absorbing pigments. Although plants absorb light primarily through the pigment chlorophyll, the light absorption function is supplemented by other accessory pigments such as carotenes and xanthophylls. This arrangement both increases the surface area for light capture and allows capture of photons with a wider range of wavelengths.

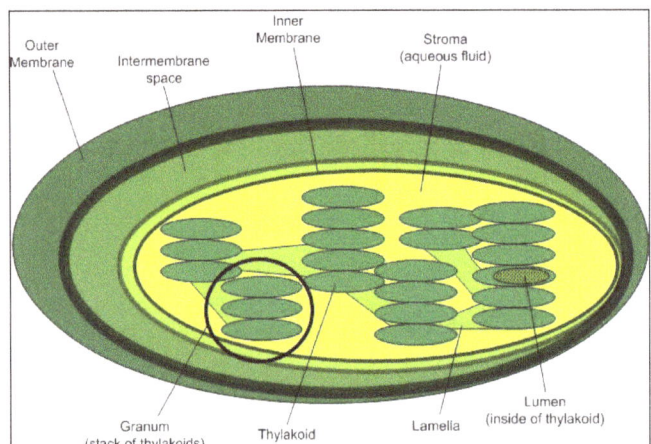

The internal structure of a chloroplast. One of the stacks of thylakoids (called a granum) is circled.

Although all cells in the green parts of a plant have chloroplasts, most light energy is captured in the leaves. The cells in the interior tissues of a leaf, called the mesophyll, can contain between 450,000 and 800,000 chloroplasts for every square millimeter of leaf. The surface of the leaf is uniformly coated with a water-resistant waxy cuticle that protects the leaf from excessive evaporation of water and decreases the absorption of ultraviolet or blue light to reduce heating.

Algae—which come in multiple forms ranging from multicellular organisms like kelp to microscopic, single-celled organisms—also contain chloroplasts and produce chlorophyll. However, various accessory pigments are also present in some algae, such as phyverdin in green algae and phycoerythrin in red algae, resulting in a wide array of colors.

Bacteria do not have Specialized Compartments for Photosynthesis

Photosynthetic bacteria do not have chloroplasts (or any membrane-bound compartments). Instead, photosynthesis takes place directly within the cell. Cyanobacteria contain thylakoid membranes very similar to those in chloroplasts and are the only prokaryotes that perform oxygen-generating photosynthesis. Other photosynthetic bacteria contain a variety of different pigments, called bacteriochlorophylls, and do not produce oxygen. Some bacteria, such as Chromatium, oxidize hydrogen sulfide instead of water, producing sulfur as a waste product.

Photosynthesis Occurs in Two Stages

The Light Reactions Convert Solar Energy to Chemical Energy

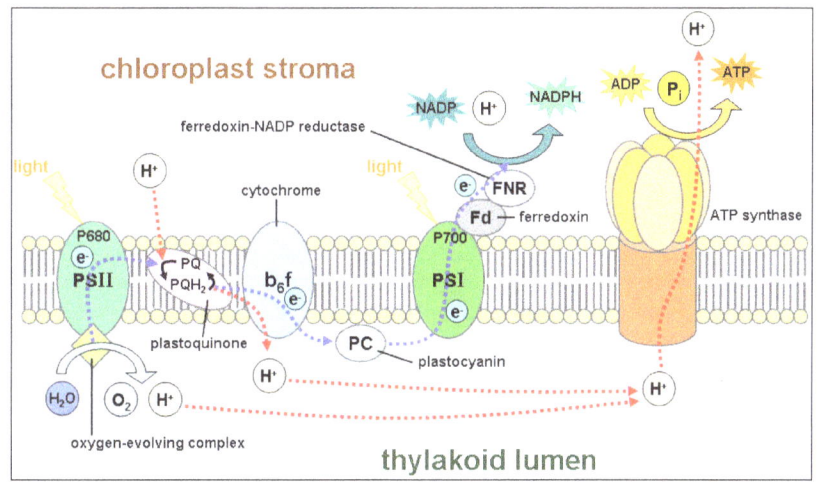

The light-dependent reactions of photosynthesis occur at the thylakoid membrane.

Photosynthesis begins when light is absorbed by chlorophyll and accessory pigments. Not all wavelengths of light can support photosynthesis. The photosynthetic action spectrum depends on the type of accessory pigments present. For example, in green plants, the chlorophylls and carotenoids absorb all visible light other than green, with peaks for violet-blue and red light. In red algae, the action spectrum overlaps with the absorption spectrum of phycobilins for blue-green light, which allows these algae to grow in deeper waters that filter out the longer wavelengths used by green plants. The non-absorbed part of the light spectrum is what gives photosynthetic organisms their color (e.g., green plants, red algae, purple bacteria) and is the least effective wavelength for photosynthesis in the respective organisms.

The electronic excitation caused by light absorption passes from one chlorophyll molecule to the next until it is trapped by a chlorophyll pair with special properties. At this site, known as the reaction center, the energy of the electron is converted into chemical energy; i.e., light is used to create a reducing potential. There are two kinds of light reactions that occur in these reaction centers, which are called photosystems:

- Photosystem I generates reducing power in the form of NADPH (a process called photoreduction).

- • Photosystem II transfers the electrons of water to a quinone (a type of aromatic compound) at the same time that it forms oxygen from the oxidation of water.

NADPH is the main reducing agent in chloroplasts, providing a source of energetic electrons to other reactions. However, its production leaves chlorophyll with a deficit of electrons, which must be obtained from some other reducing agent. The source of these electrons in green-plant and cyanobacterial photosynthesis is water.

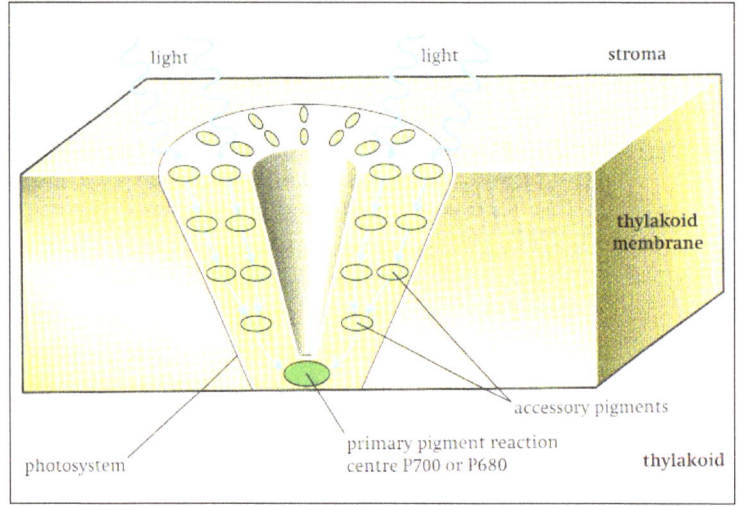

A light-harvesting cluster of photosynthetic pigments (called a photosystem)
is present in the thylakoid membrane of chloroplasts.

Electron flow within and between each photosystem generates a transmembrane proton gradient that drives the synthesis of ATP, through a process known as photophosphorylation. When a chlorophyll molecule at the core of the photosystem II reaction center obtains sufficient excitation energy from the adjacent antenna pigments, an electron is transferred to the primary electron-acceptor molecule through a process called photoinduced charge separation. These electrons are shuttled through an electron transport chain, the Z-scheme shown in the diagram, that initially functions to generate a chemiosmotic potential across the membrane. An ATP synthase enzyme uses the chemiosmotic potential to make ATP, while NADPH is a product of the terminal redox reaction.

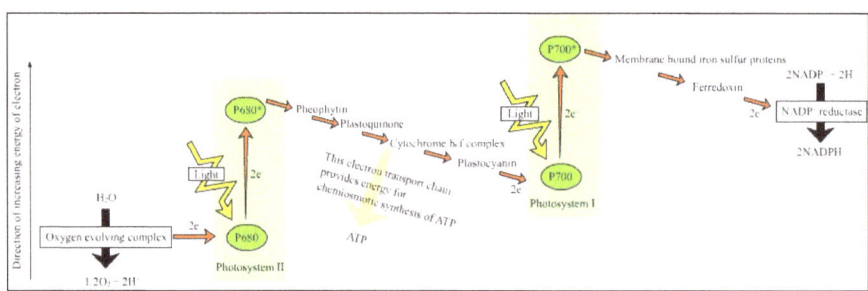

The Z-scheme is an electron transport chain that generates the chemioosmotic potential used to
synthesize ATP. It is so-called because the redox diagram takes the shape of a Z.

The pathway described above is referred to as non-cyclic photophosphorylation. However, an alternative pathway is cyclic photophosphorylation, in which ATP is generated without the concomitant formation of NADPH. This pathway is utilized when NAD^+ is unavailable to accept electrons.

The cyclic reaction takes place only at photosystem I. Once the electron is displaced, it is passed down the electron acceptor molecules and returns to photosystem I.

Synthesis of Organic Compounds

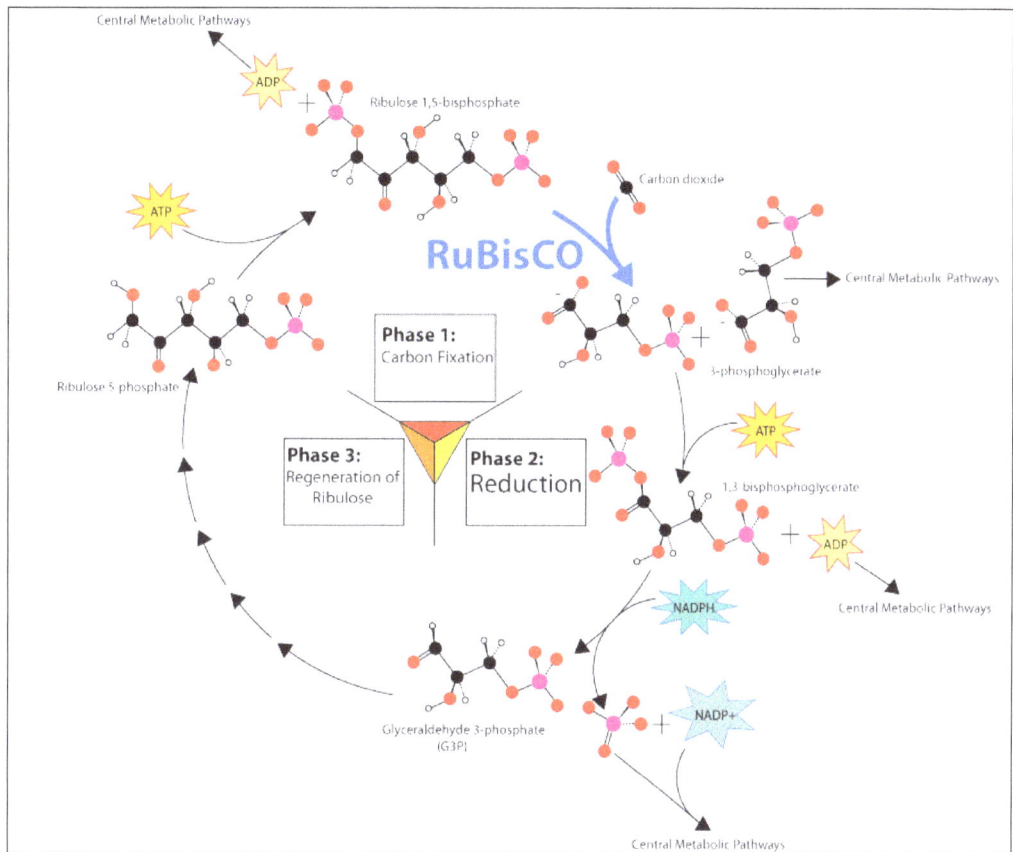

Overview of the Calvin cycle and carbon fixation.

Plants use chemical energy generated from ATP and NADPH to fix carbon dioxide (a process also known as carbon reduction) into carbohydrates and other organic compounds through light-independent reactions (or the Calvin cycle). They reduce carbon dioxide and convert it into 3-phosphoglycerate in a series of reactions that occur in the stroma (the fluid-filled interior) of the chloroplast. Hexoses (six-carbon sugars) such as glucose are then formed from 3-phosphoglycerate by the gluconeogenic pathway.

Specifically, the fixation of carbon dioxide is a light-independent process in which carbon dioxide combines with a five-carbon sugar, ribulose 1,5-bisphosphate (RuBP), to form a six-carbon compound. This compound is hydrolyzed to two molecules of a three-carbon compound, glycerate 3-phosphate (GP), also known as 3-phosphoglycerate (PGA). In the presence of ATP and NADPH from the light-dependent stages, GP is reduced to glyceraldehyde 3-phosphate (G3P). This product is also referred to as 3-phosphoglyceraldehyde (PGAL) or even as triose phosphate (where triose refers to a 3-carbon sugar). This reaction is catalyzed by an enzyme commonly called rubisco (after ribulose 1,5-bisphosphate carboxylase/oxygenase), located on the stromal surface of the thylakoid membrane. Rubisco is the most abundant enzyme, and probably the most abundant protein, in the biosphere, accounting for more than sixteen percent of the total protein of chloroplasts.

Five out of six molecules of the G3P produced are used to regenerate the enzyme RuBP, so that the process can continue. One out of six molecules of the triose phosphates not "recycled" often condenses to form hexose phosphate, which ultimately yields sucrose, starch and cellulose. The sugars produced during carbon metabolism yield carbon skeletons that can be used for other metabolic reactions like the production of amino acids and lipids.

Three molecules of ATP and 2 molecules of NADPH are consumed in converting carbon dioxide into a one molecule of a hexose such as glucose or fructose.

Alternative Methods of Carbon Fixation have Evolved to Meet Environmental Conditions

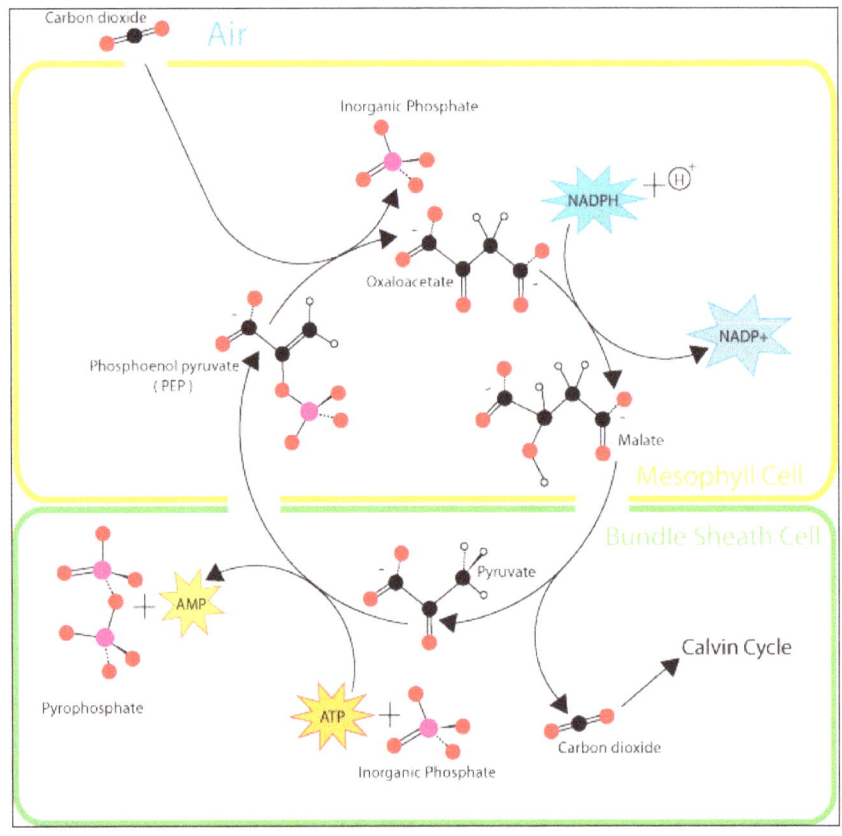

Overview of C_4 carbon fixation.

In hot and dry conditions, plants will close their stomata (small openings on the underside of leaves used for gas exchange) to prevent loss of water. Under these conditions, oxygen gas, produced by the light reactions of photosynthesis, will concentrate in the leaves, causing photorespiration to occur. Photorespiration is a wasteful reaction: organic carbon is converted into carbon dioxide without the production of ATP, NADPH, or another energy-rich metabolite.

Rubisco, the enzyme that captures carbon dioxide in the light-independent reactions, has a binding affinity for both carbon dioxide and oxygen. When the concentration of carbon dioxide is high, rubisco will fix carbon dioxide. However, if the oxygen concentration is high, rubisco will bind oxygen instead of carbon dioxide. Rubisco's tendency to catalyze this oxygenase activity increases more rapidly with temperature than its carboxylase activity.

The solution arrived at by the C_4 plants (which include many important crop plants such as maize, sorghum, sugarcane, and millet) is to achieve a high concentration of carbon dioxide in the leaves (the site of the Calvin cycle) under these conditions.

C_4 plants capture carbon dioxide using an enzyme called PEP carboxylase that adds carbon dioxide to the 3-carbon molecule phosphoenolpyruvate (PEP), creating the 4-carbon molecule oxaloacetic acid. Plants without this enzyme are called C_3 plants because the primary carboxylation reaction produces the 3-carbon sugar 3-phosphoglycerate directly in the Calvin cycle. When oxygen levels rise in the leaf, C_4 plants plants reverse the reaction to release carbon dioxide, thus preventing photorespiration. Through this mechanism, C_4 plants can produce more sugar than C_3 plants in conditions of strong light and high temperature. These C_4 plants compounds carry carbon dioxide from mesophyll cells, which are in contact with air, to bundle-sheath cells, which are major sites of photosynthesis.

The pineapple is an example of a CAM plant.

Plants living in arid conditions, such as cacti and most succulents, can also use PEP carboxylase to capture carbon dioxide in a process called Crassulacean acid metabolism (CAM). CAM plants close their stomata during the day in order to conserve water by preventing evapotranspiration. Their stomata then open during the cooler and more humid nighttime hours, allowing uptake of carbon dioxide for use in carbon fixation. By thus reducing evapotranspiration rates during gas exchange, CAM allows plants to grow in environments that would otherwise be far too dry for plant growth or, at best, would subject them to severe drought stress. Although they resemble C_4 plants in some respects, CAM plants store the CO_2 in different molecules and have a different leaf anatomy than C_4 plants.

In sum, C_4 plants metabolism physically separates CO_2 fixation from the Calvin cycle, while CAM metabolism temporally separates CO_2 fixation from the Calvin cycle.

Light-dependent Reactions

In photosynthesis, the light-dependent reactions take place on the thylakoid membranes. The inside of the thylakoid membrane is called the lumen, and outside the thylakoid membrane is the stroma, where the light-independent reactions take place. The thylakoid membrane contains some integral membrane protein complexes that catalyze the light reactions. There are four major protein complexes in the thylakoid membrane: Photosystem II (PSII), Cytochrome b6f complex, Photosystem I (PSI), and ATP synthase. These four complexes work together to ultimately create the products ATP and NADPH.

The four photosystems absorb light energy through pigments—primarily the chlorophylls, which are responsible for the green color of leaves. The light-dependent reactions begin in photosystem II. When a chlorophyll a molecule within the reaction center of PSII absorbs a photon, an electron in this molecule attains a higher energy level. Because this state of an electron is very unstable, the electron is transferred from one to another molecule creating a chain of redox reactions, called an electron transport chain (ETC). The electron flow goes from PSII to cytochrome b_6f to PSI. In PSI, the electron gets the energy from another photon. The final electron acceptor is NADP. In oxygenic photosynthesis, the first electron donor is water, creating oxygen as a waste product. In anoxygenic photosynthesis various electron donors are used.

Cytochrome b_6f and ATP synthase work together to create ATP. This process is called photophosphorylation, which occurs in two different ways. In non-cyclic photophosphorylation, cytochrome b_6f uses the energy of electrons from PSII to pump protons from the stroma to the lumen. The proton gradient across the thylakoid membrane creates a proton-motive force, used by ATP synthase to form ATP. In cyclic photophosphorylation, cytochrome b_6f uses the energy of electrons from not only PSII but also PSI to create more ATP and to stop the production of NADPH. Cyclic phosphorylation is important to create ATP and maintain NADPH in the right proportion for the light-independent reactions.

The net-reaction of all light-dependent reactions in oxygenic photosynthesis is:

$$2H_2O + 2NADP^+ + 3ADP + 3P_i \rightarrow O_2 + 2NADPH + 3ATP$$

The two photosystems are protein complexes that absorb photons and are able to use this energy to create a photosynthetic electron transport chain. Photosystem I and II are very similar in structure and function. They use special proteins, called light-harvesting complexes, to absorb the photons with very high effectiveness. If a special pigment molecule in a photosynthetic reaction center absorbs a photon, an electron in this pigment attains the excited state and then is transferred to another molecule in the reaction center. This reaction, called photoinduced charge separation, is the start of the electron flow and is unique because it transforms light energy into chemical forms.

In chemistry, many reactions depend on the absorption of photons to provide the energy needed to overcome the activation energy barrier and hence can be labelled light-dependent. Such reactions range from the silver halide reactions used in photographic film to the creation and destruction of ozone in the upper atmosphere.

The reaction center is in the thylakoid membrane. It transfers light energy to a dimer of chlorophyll pigment molecules near the periplasmic (or thylakoid lumen) side of the membrane. This dimer is called a special pair because of its fundamental role in photosynthesis. This special pair is slightly different in PSI and PSII reaction center. In PSII, it absorbs photons with a wavelength of 680 nm, and it is therefore called P680. In PSI, it absorbs photons at 700 nm, and it is called P700. In bacteria, the special pair is called P760, P840, P870, or P960. "P" here means pigment, and the number following it is the wavelength of light absorbed.

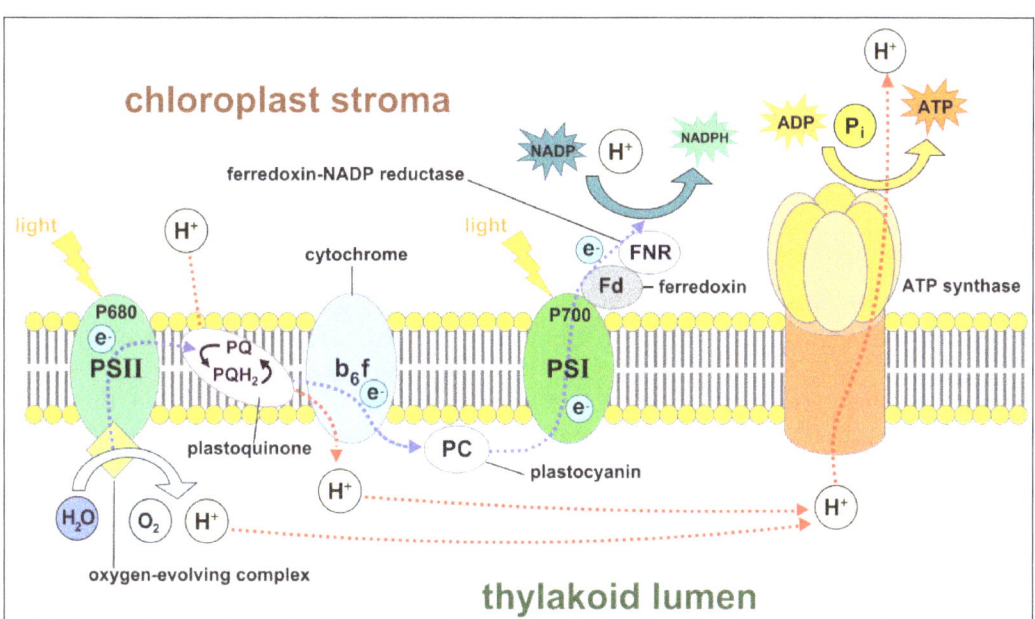

Light-dependent reactions of photosynthesis at the thylakoid membrane.

If an electron of the special pair in the reaction center becomes excited, it cannot transfer this energy to another pigment using resonance energy transfer. In normal circumstances, the electron should return to the ground state, but, because the reaction center is arranged so that a suitable electron acceptor is nearby, the excited electron can move from the initial molecule to the acceptor. This process results in the formation of a positive charge on the special pair (due to the loss of an electron) and a negative charge on the acceptor and is, hence, referred to as photoinduced charge separation. In other words, electrons in pigment molecules can exist at specific energy levels. Under normal circumstances, they exist at the lowest possible energy level they can. However, if there is enough energy to move them into the next energy level, they can absorb that energy and occupy that higher energy level. The light they absorb contains the necessary amount of energy needed to push them into the next level. Any light that does not have enough or has too much energy cannot be absorbed and is reflected. The electron in the higher energy level, however, does not want to be there; the electron is unstable and must return to its normal lower energy level. To do this, it must release the energy that has put it into the higher energy state to begin with. This can happen various ways. The extra energy can be converted into molecular motion and lost as heat. Some of the extra energy can be lost as heat energy, while the rest is lost as light. (This re-emission of light energy is called fluorescence.) The energy, but not the e- itself, can be passed onto another molecule. (This is called resonance.) The energy and the e- can be transferred to another molecule. Plant pigments usually utilize the last two of these reactions to convert the sun's energy into their own.

This initial charge separation occurs in less than 10 picoseconds (10^{-11} seconds). In their high-energy states, the special pigment and the acceptor could undergo charge recombination; that is, the electron on the acceptor could move back to neutralize the positive charge on the special pair. Its return to the special pair would waste a valuable high-energy electron and simply convert the absorbed light energy into heat. In the case of PSII, this backflow of electrons can produce reactive oxygen species leading to photoinhibition. Three factors in the structure of the reaction center work together to suppress charge recombination nearly completely.

- Another electron acceptor is less than 10 Å away from the first acceptor, and so the electron is rapidly transferred farther away from the special pair.

- An electron donor is less than 10 Å away from the special pair, and so the positive charge is neutralized by the transfer of another electron.

- The electron transfer back from the electron acceptor to the positively charged special pair is especially slow. The rate of an electron transfer reaction increases with its thermodynamic favorability up to a point and then decreases. The back transfer is so favourable that it takes place in the inverted region where electron-transfer rates become slower.

Thus, electron transfer proceeds efficiently from the first electron acceptor to the next, creating an electron transport chain that ends if it has reached NADPH.

In Chloroplasts

The photosynthesis process in chloroplasts begins when an electron of P680 of PSII attains a higher-energy level. This energy is used to reduce a chain of electron acceptors that have subsequently lowered redox-potentials. This chain of electron acceptors is known as an electron transport chain. When this chain reaches PS I, an electron is again excited, creating a high redox-potential. The electron transport chain of photosynthesis is often put in a diagram called the z-scheme, because the redox diagram from P680 to P700 resembles the letter Z.

The final product of PSII is plastoquinol, a mobile electron carrier in the membrane. Plastoquinol transfers the electron from PSII to the proton pump, cytochrome b6f. The ultimate electron donor of PSII is water. Cytochrome b_6f proceeds the electron chain to PSI through plastocyanin molecules. PSI is able to continue the electron transfer in two different ways. It can transfer the electrons either to plastoquinol again, creating a cyclic electron flow, or to an enzyme called FNR (Ferredoxin—NADP(+) reductase), creating a non-cyclic electron flow. PSI releases FNR into the stroma, where it reduces $NADP^+$ to NADPH.

Activities of the electron transport chain, especially from cytochrome b_6f, lead to pumping of protons from the stroma to the lumen. The resulting transmembrane proton gradient is used to make ATP via ATP synthase.

The overall process of the photosynthetic electron transport chain in chloroplasts is:

$$H_2O \rightarrow PS\ II \rightarrow plastoquinone \rightarrow cyt\ b_6f \rightarrow plastocyanin \rightarrow PS\ I \rightarrow NADPH$$

Photosystem II

PS II is extremely complex, a highly organized transmembrane structure that contains a water-splitting complex, chlorophylls and carotenoid pigments, a reaction center (P680), pheophytin (a pigment similar to chlorophyll), and two quinones. It uses the energy of sunlight to transfer electrons from water to a mobile electron carrier in the membrane called plastoquinone:

$$H_2O \rightarrow P680 \rightarrow P680^* \rightarrow plastoquinone$$

Plastoquinone, in turn, transfers electrons to cyt b_6f, which feeds them into PS I.

The Water-Splitting Complex

The step $H_2O \rightarrow P680$ is performed by a poorly understood structure embedded within PS II called the water-splitting complex or the oxygen-evolving complex. It catalyzes a reaction that splits water into electrons, protons and oxygen:

$$2H_2O \rightarrow 4H^+ + 4e^- + O_2$$

The actual steps of the above reaction are running in the following way (Dolai's diagram of S-states): (I) $2H_2O$ (monoxide) (II) $OH. H_2O$ (hydroxide) (III) H_2O_2 (peroxide) (IV) HO_2 (super oxide) (V) O_2 (di-oxygen). (Dolai's mechanism)

The electrons are transferred to special chlorophyll molecules (embedded in PS II) that are promoted to a higher-energy state by the energy of photons.

The Reaction Center

The excitation $P680 \rightarrow P680^*$ of the reaction center pigment $P680$ occurs here. These special chlorophyll molecules embedded in PS II absorb the energy of photons, with maximal absorption at 680 nm. Electrons within these molecules are promoted to a higher-energy state. This is one of two core processes in photosynthesis, and it occurs with astonishing efficiency (greater than 90%) because, in addition to direct excitation by light at 680 nm, the energy of light first harvested by antenna proteins at other wavelengths in the light-harvesting system is also transferred to these special chlorophyll molecules.

This is followed by the step $P680^* \rightarrow$ pheophytin, and then on to plastoquinone, which occurs within the reaction center of PS II. High-energy electrons are transferred to plastoquinone before it subsequently picks up two protons to become plastoquinol. Plastoquinol is then released into the membrane as a mobile electron carrier.

This is the second core process in photosynthesis. The initial stages occur within picoseconds, with an efficiency of 100%. The seemingly impossible efficiency is due to the precise positioning of molecules within the reaction center. This is a solid-state process, not a chemical reaction. It occurs within an essentially crystalline environment created by the macromolecular structure of PS II. The usual rules of chemistry (which involve random collisions and random energy distributions) do not apply in solid-state environments.

Link of Water-Splitting Complex and Chlorophyll Excitation

When the chlorophyll passes the electron to pheophytin, it obtains an electron from P_{680}^*. In turn, P_{680}^* can oxidize the Z (or Y_Z) molecule. Once oxidized, the Z molecule can derive electrons from the oxygen-evolving complex (OEC). Dolai's S-state diagrams show the reactions of water splitting in the oxygen-evolving complex.

Cytochrome b_6f

PS II and PS I are connected by a transmembrane proton pump, cytochrome b_6f complex (plastoquinol—plastocyanin reductase; EC 1.10.99.1). Electrons from PS II are carried by plastoquinol to cyt b_6f, where they are removed in a stepwise fashion (reforming plastoquinone) and transferred to a water-soluble electron carrier called plastocyanin. This redox process is coupled to the pumping of four protons across the membrane. The resulting proton gradient (together with the proton gradient produced by the water-splitting complex in PS II) is used to make ATP via ATP synthase.

The structure and function of cytochrome b_6f (in chloroplasts) is very similar to cytochrome bc_1 (Complex III in mitochondria). Both are transmembrane structures that remove electrons from a mobile, lipid-soluble electron carrier (plastoquinone in chloroplasts; ubiquinone in mitochondria) and transfer them to a mobile, water-soluble electron carrier (plastocyanin in chloroplasts; cytochrome c in mitochondria). Both are proton pumps that produce a transmembrane proton gradient. In fact, cytochrome b_6 and subunit IV are homologous to mitochondrial cytochrome b and the Rieske iron-sulfur proteins of the two complexes are homologous. However, cytochrome f and cytochrome c_1 are not homologous.

Photosystem I

PS I accepts electrons from plastocyanin and transfers them either to NADPH (noncyclic electron transport) or back to cytochrome b_6f (cyclic electron transport):

$$\text{plastocyanin} \rightarrow \text{P700} \rightarrow \text{P700}^* \rightarrow \text{FNR} \rightarrow \text{NADPH}$$
$$\uparrow \qquad\qquad\qquad \downarrow$$
$$b_6f \qquad \leftarrow \quad \text{Plastoquinone}$$

PS I, like PS II, is a complex, highly organized transmembrane structure that contains antenna chlorophylls, a reaction center (P700), phylloquinine, and a number of iron-sulfur proteins that serve as intermediate redox carriers.

The light-harvesting system of PS I uses multiple copies of the same transmembrane proteins used by PS II. The energy of absorbed light (in the form of delocalized, high-energy electrons) is funneled into the reaction center, where it excites special chlorophyll molecules (P700, maximum light absorption at 700 nm) to a higher energy level. The process occurs with astonishingly high efficiency.

Electrons are removed from excited chlorophyll molecules and transferred through a series of intermediate carriers to ferredoxin, a water-soluble electron carrier. As in PS II, this is a solid-state process that operates with 100% efficiency.

There are two different pathways of electron transport in PS I. In noncyclic electron transport, ferredoxin carries the electron to the enzyme ferredoxin $NADP^+$ reductase (FNR) that reduces $NADP^+$ to NADPH. In cyclic electron transport, electrons from ferredoxin are transferred (via plastoquinone) to a proton pump, cytochrome b_6f. They are then returned (via plastocyanin) to P700.

NADPH and ATP are used to synthesize organic molecules from CO_2. The ratio of NADPH to ATP production can be adjusted by adjusting the balance between cyclic and noncyclic electron transport.

It is noteworthy that PS I closely resembles photosynthetic structures found in green sulfur bacteria, just as PS II resembles structures found in purple bacteria.

In Bacteria

PS II, PS I, and cytochrome b_6f are found in chloroplasts. All plants and all photosynthetic algae contain chloroplasts, which produce NADPH and ATP by the mechanisms described above. In essence, the same transmembrane structures are also found in cyanobacteria.

Unlike plants and algae, cyanobacteria are prokaryotes. They do not contain chloroplasts. Rather, they bear a striking resemblance to chloroplasts themselves. This suggests that organisms resembling cyanobacteria were the evolutionary precursors of chloroplasts. One imagines primitive eukaryotic cells taking up cyanobacteria as intracellular symbionts in a process known as endosymbiosis.

Cyanobacteria

Cyanobacteria contain both PS I and PS II. Their light-harvesting system is different from that found in plants (they use phycobilins, rather than chlorophylls, as antenna pigments), but their electron transport chain:

$$H_2O \rightarrow PS\ II \rightarrow plastoquinone \rightarrow b_6f \rightarrow cytochrome c_6 \rightarrow PS\ I \rightarrow ferredoxin \rightarrow NADPH$$

$$\uparrow \qquad\qquad\qquad\qquad \downarrow$$
$$b_6f \qquad \leftarrow \qquad plastoquinone$$

is, in essence, the same as the electron transport chain in chloroplasts. The mobile water-soluble electron carrier is cytochrome c_6 in cyanobacteria, plastocyanin in plants.

Cyanobacteria can also synthesize ATP by oxidative phosphorylation, in the manner of other bacteria. The electron transport chain is

$$NADH\ dehydrogenase \rightarrow plastoquinone \rightarrow b_6f \rightarrow cyt\ c_6 \rightarrow cyt\ aa_3 \rightarrow O_2$$

where the mobile electron carriers are plastoquinone and cytochrome c_6, while the proton pumps are NADH dehydrogenase, cyt b_6f and cytochrome aa_3 (member of the COX3 family).

Cyanobacteria are the only bacteria that produce oxygen during photosynthesis. Earth's primordial atmosphere was anoxic. Organisms like cyanobacteria produced our present-day oxygen-containing atmosphere.

The other two major groups of photosynthetic bacteria, purple bacteria and green sulfur bacteria, contain only a single photosystem and do not produce oxygen.

Purple Bacteria

Purple bacteria contain a single photosystem that is structurally related to PS II in cyanobacteria and chloroplasts:

$$P870 \rightarrow P870^* \rightarrow ubiquinone \rightarrow cyt\ bc_1 \rightarrow cyt\ c_2 \rightarrow P870$$

This is a cyclic process in which electrons are removed from an excited chlorophyll molecule (bacteriochlorophyll; P870), passed through an electron transport chain to a proton pump (cytochrome bc_1 complex; similar to the chloroplastic one), and then returned to the chlorophyll molecule. The result is a proton gradient, which is used to make ATP via ATP synthase. As in cyanobacteria and chloroplasts, this is a solid-state process that depends on the precise orientation of various functional groups within a complex transmembrane macromolecular structure.

To make NADPH, purple bacteria use an external electron donor (hydrogen, hydrogen sulfide, sulfur, sulfite, or organic molecules such as succinate and lactate) to feed electrons into a reverse electron transport chain.

Green Sulfur Bacteria

Green sulfur bacteria contain a photosystem that is analogous to PS I in chloroplasts:

$$P840 \rightarrow P840^* \rightarrow ferredoxin \rightarrow NADH$$
$$\uparrow \qquad\qquad\qquad \downarrow$$
$$cyt\ c_{553} \leftarrow bc_1 \leftarrow menaquinone$$

There are two pathways of electron transfer. In cyclic electron transfer, electrons are removed from an excited chlorophyll molecule, passed through an electron transport chain to a proton pump, and then returned to the chlorophyll. The mobile electron carriers are, as usual, a lipid-soluble quinone and a water-soluble cytochrome. The resulting proton gradient is used to make ATP.

In noncyclic electron transfer, electrons are removed from an excited chlorophyll molecule and used to reduce NAD^+ to NADH. The electrons removed from P840 must be replaced. This is accomplished by removing electrons from H_2S, which is oxidized to sulfur (hence the name "green sulfur bacteria").

Purple bacteria and green sulfur bacteria occupy relatively minor ecological niches in the present day biosphere. They are of interest because of their importance in precambrian ecologies, and because their methods of photosynthesis were the likely evolutionary precursors of those in modern plants.

Light-independent Reactions

The light-independent reactions of photosynthesis are the chemical reactions that convert carbon dioxide and other compounds into glucose. These reactions occur in the stroma, the fluid-filled area of a chloroplast outside the thylakoid membranes. These reactions take the products (ATP and NADPH) of light-dependent reactions and perform further chemical processes on them. There are three phases to the light-independent reactions, collectively called the Calvin cycle: carbon fixation, reduction reactions, and ribulose 1,5-bisphosphate (RuBP) regeneration.

Though it is called the "dark reactions", the Calvin cycle does not actually occur in the dark or during nighttime. This is because the process requires reduced NADP which is short-lived and comes from the light-dependent reactions. In the dark, plants instead release sucrose into the phloem from their starch reserves to provide energy for the plant. The Calvin cycle thus happens when light is available independent of the kind of photosynthesis (C3 carbon fixation, C4 carbon fixation, and Crassulacean Acid Metabolism (CAM)); CAM plants store malic acid in their vacuoles every night and release it by day to make this process work.

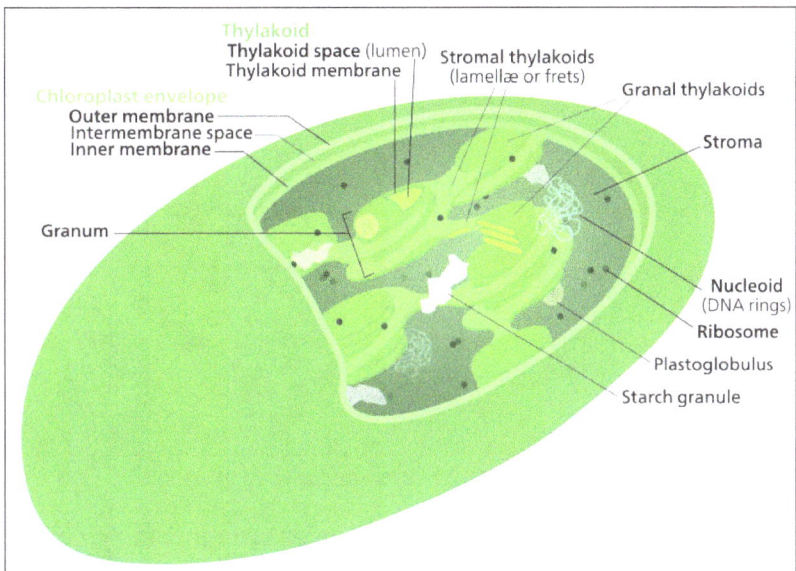

The internal structure of a chloroplast.

Coupling to other Metabolic Pathways

These reactions are closely coupled to the thylakoid electron transport chain as the energy required to reduce the carbon dioxide is provided by NADPH produced in photosystem I during the light dependent reactions. The process of photorespiration, also known as C2 cycle, is also coupled to the calvin cycle, as it results from an alternative reaction of the RuBisCO enzyme, and its final byproduct is another glyceraldehyde-3-P.

Calvin Cycle

The Calvin cycle, Calvin–Benson–Bassham (CBB) cycle, reductive pentose phosphate cycle or C3 cycle is a series of biochemical redox reactions that take place in the stroma of chloroplast in photosynthetic organisms.

The cycle was discovered by Melvin Calvin, James Bassham, and Andrew Benson at the University of California, Berkeley by using the radioactive isotope carbon-14.

Photosynthesis occurs in two stages in a cell. In the first stage, light-dependent reactions capture the energy of light and use it to make the energy-storage and transport molecules ATP and NADPH. The Calvin cycle uses the energy from short-lived electronically excited carriers to convert carbon dioxide and water into organic compounds that can be used by the organism (and by animals that feed on it). This set of reactions is also called carbon fixation. The key enzyme of the cycle is called RuBisCO. In the following biochemical equations, the chemical species (phosphates and carboxylic acids) exist in equilibria among their various ionized states as governed by the pH.

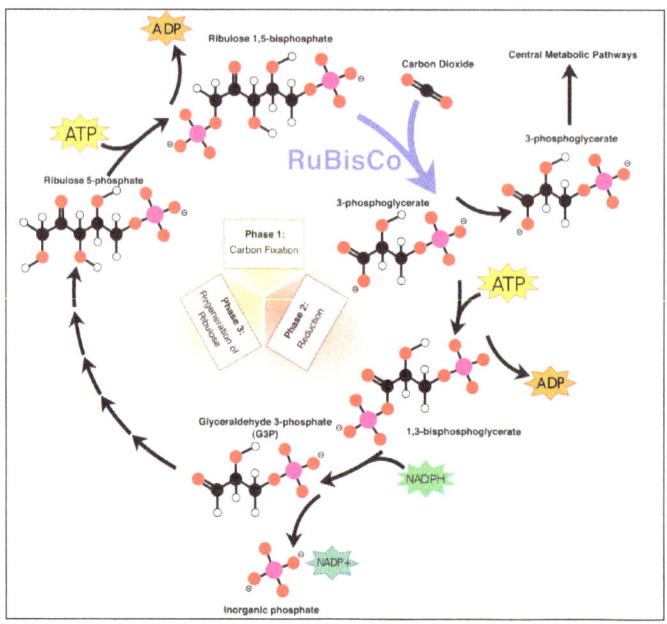

Overview of the Calvin cycle and carbon fixation.

The enzymes in the Calvin cycle are functionally equivalent to most enzymes used in other metabolic pathways such as gluconeogenesis and the pentose phosphate pathway, but they are found in the chloroplast stroma instead of the cell cytosol, separating the reactions. They are activated in the light (which is why the name "dark reaction" is misleading), and also by products of the light-dependent reaction. These regulatory functions prevent the Calvin cycle from being respired to carbon dioxide. Energy (in the form of ATP) would be wasted in carrying out these reactions that have no net productivity.

The sum of reactions in the Calvin cycle is the following:

$3\ CO_2 + 6\ NADPH + 6\ H+ + 9\ ATP \rightarrow$ glyceraldehyde-3-phosphate (G3P) $+ 6\ NADP+ + 9\ ADP + 3\ H_2O + 8\ Pi$ (Pi = inorganic phosphate)

Hexose (six-carbon) sugars are not a product of the Calvin cycle. Although many texts list a product of photosynthesis as $C_6H_{12}O_6$, this is mainly a convenience to counter the equation of respiration, where six-carbon sugars are oxidized in mitochondria. The carbohydrate products of the Calvin cycle are three-carbon sugar phosphate molecules, or "triose phosphates", namely, glyceraldehyde-3-phosphate (G3P).

Steps

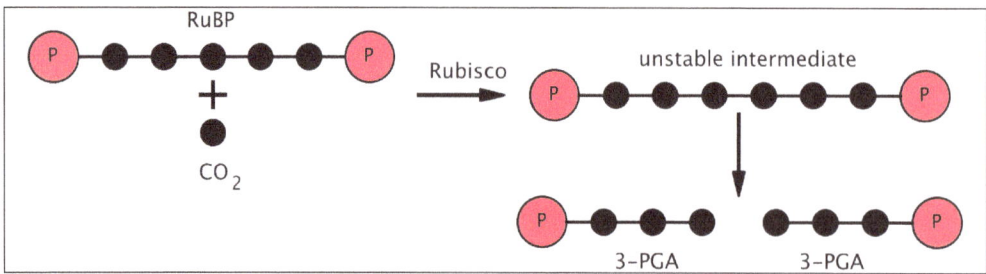

Calvin cycle step 1 (black circles represent carbon atoms).

In the first stage of the Calvin cycle, a CO_2 molecule is incorporated into one of two three-carbon molecules (glyceraldehyde 3-phosphate or G3P), where it uses up two molecules of ATP and two molecules of NADPH, which had been produced in the light-dependent stage. The three steps involved are:

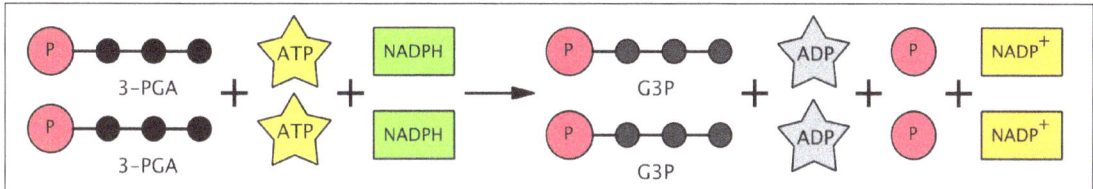

Calvin cycle steps 2 and 3 combined.

- The enzyme RuBisCO catalyses the carboxylation of ribulose-1,5-bisphosphate, RuBP, a 5-carbon compound, by carbon dioxide (a total of 6 carbons) in a two-step reaction. The product of the first step is enediol-enzyme complex that can capture CO_2 or O_2. Thus, enediol-enzyme complex is the real carboxylase/oxygenase. The CO_2 that is captured by enediol in second step produces an unstable six-carbon compound called 2-carboxy 3-keto 1,5-biphosphoribotol (or 3-keto-2-carboxyarabinitol 1,5-bisphosphate) that immediately splits into 2 molecules of 3-phosphoglycerate, or 3-PGA, a 3-carbon compound (also: 3-phosphoglyceric acid, PGA, 3PGA).

- The enzyme phosphoglycerate kinase catalyses the phosphorylation of 3-PGA by ATP (which was produced in the light-dependent stage). 1,3-Bisphosphoglycerate (1,3BPGA, glycerate-1,3-bisphosphate) and ADP are the products. (However, note that two 3-PGAs are produced for every CO_2 that enters the cycle, so this step utilizes two ATP per CO_2 fixed.)

- The enzyme glyceraldehyde 3-phosphate dehydrogenase catalyses the reduction of 1,3BPGA by NADPH (which is another product of the light-dependent stage). Glyceraldehyde 3-phosphate (also called G3P, GP, TP, PGAL, GAP) is produced, and the NADPH itself is oxidized and becomes $NADP^+$. Again, two NADPH are utilized per CO_2 fixed.

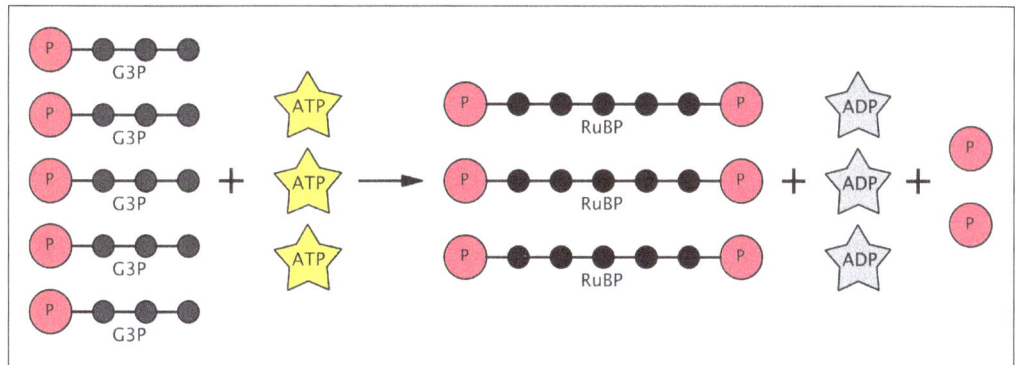

Regeneration stage of the Calvin cycle.

The next stage in the Calvin cycle is to regenerate RuBP. Five G3P molecules produce three RuBP molecules, using up three molecules of ATP. Since each CO_2 molecule produces two G3P molecules, three CO_2 molecules produce six G3P molecules, of which five are used to regenerate RuBP, leaving a net gain of one G3P molecule per three CO_2 molecules (as would be expected from the number of carbon atoms involved).

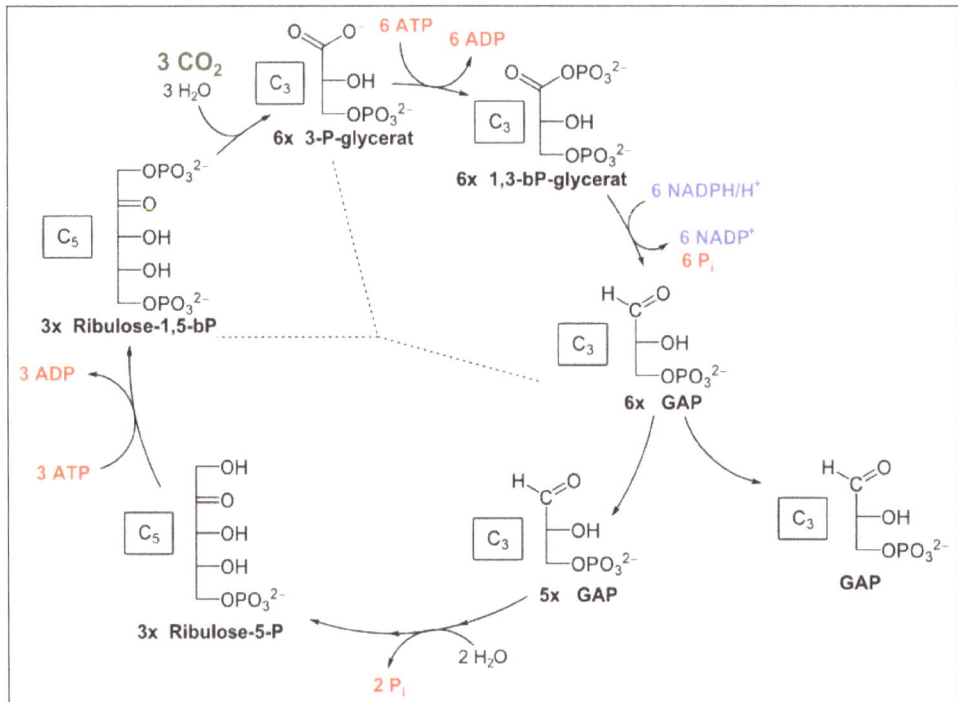

Simplified C3 cycle with structural formulas.

The regeneration stage can be broken down into steps.

- Triose phosphate isomerase converts all of the G3P reversibly into dihydroxyacetone phosphate (DHAP), also a 3-carbon molecule.

- Aldolase and fructose-1,6-bisphosphatase convert a G3P and a DHAP into fructose 6-phosphate (6C). A phosphate ion is lost into solution.

- Then fixation of another CO_2 generates two more G3P.

- F6P has two carbons removed by transketolase, giving erythrose-4-phosphate. The two carbons on transketolase are added to a G3P, giving the ketose xylulose-5-phosphate (Xu5P).

- E4P and a DHAP (formed from one of the G3P from the second CO_2 fixation) are converted into sedoheptulose-1,7-bisphosphate (7C) by aldolase enzyme.

- Sedoheptulose-1,7-bisphosphatase (one of only three enzymes of the Calvin cycle that are unique to plants) cleaves sedoheptulose-1,7-bisphosphate into sedoheptulose-7-phosphate, releasing an inorganic phosphate ion into solution.

- Fixation of a third CO_2 generates two more G3P. The ketose S7P has two carbons removed by transketolase, giving ribose-5-phosphate (R5P), and the two carbons remaining on transketolase are transferred to one of the G3P, giving another Xu5P. This leaves one G3P as the product of fixation of 3 CO_2, with generation of three pentoses that can be converted to Ru5P.

- R5P is converted into ribulose-5-phosphate (Ru5P, RuP) by phosphopentose isomerase. Xu5P is converted into RuP by phosphopentose epimerase.

- Finally, phosphoribulokinase (another plant-unique enzyme of the pathway) phosphory- lates RuP into RuBP, ribulose-1,5-bisphosphate, completing the Calvin cycle. This requires the input of one ATP.

Thus, of six G3P produced, five are used to make three RuBP (5C) molecules (totaling 15 carbons), with only one G3P available for subsequent conversion to hexose. This requires nine ATP mole- cules and six NADPH molecules per three CO_2 molecules. The equation of the overall Calvin cycle is shown diagrammatically below:

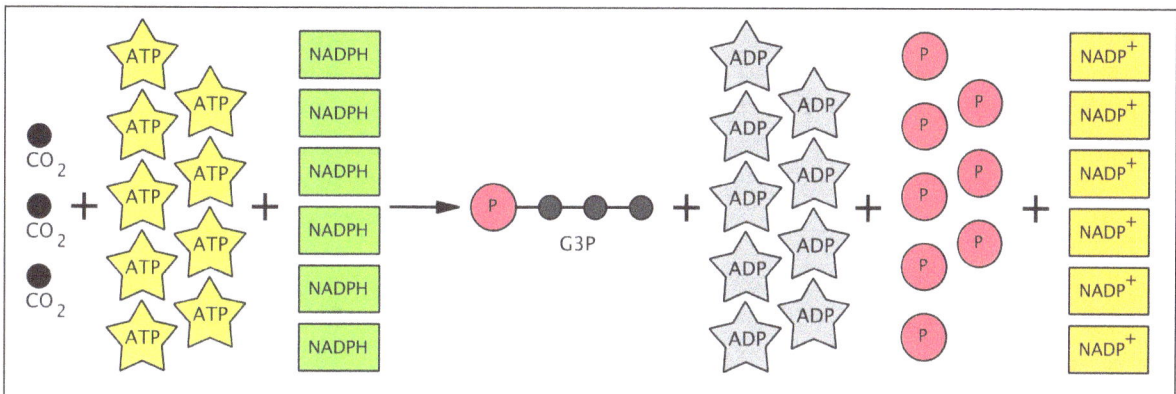

The overall equation of the Calvin cycle (black circles represent carbon atoms).

RuBisCO also reacts competitively with O_2 instead of CO_2 in photorespiration. The rate of photo- respiration is higher at high temperatures. Photorespiration turns RuBP into 3-PGA and 2-phos- phoglycolate, a 2-carbon molecule that can be converted via glycolate and glyoxalate to glycine. Via the glycine cleavage system and tetrahydrofolate, two glycines are converted into serine $+CO_2$. Serine can be converted back to 3-phosphoglycerate. Thus, only 3 of 4 carbons from two phospho- glycolates can be converted back to 3-PGA. It can be seen that photorespiration has very negative consequences for the plant, because, rather than fixing CO_2, this process leads to loss of CO_2. C4 carbon fixation evolved to circumvent photorespiration, but can occur only in certain plants native to very warm or tropical climates—corn, for example.

Products

The immediate products of one turn of the Calvin cycle are 2 glyceraldehyde-3-phosphate (G3P) molecules, 3 ADP, and 2 $NADP^+$. (ADP and $NADP^+$ are not really "products." They are regener- ated and later used again in the Light-dependent reactions). Each G3P molecule is composed of 3 carbons. For the Calvin cycle to continue, RuBP (ribulose 1,5-bisphosphate) must be regen- erated. So, 5 out of 6 carbons from the 2 G3P molecules are used for this purpose. Therefore, there is only 1 net carbon produced to play with for each turn. To create 1 surplus G3P requires 3 carbons, and therefore 3 turns of the Calvin cycle. To make one glucose molecule (which can be created from 2 G3P molecules) would require 6 turns of the Calvin cycle. Surplus G3P can also be used to form other carbohydrates such as starch, sucrose, and cellulose, depending on what the plant needs.

Light-dependent Regulation

These reactions do not occur in the dark or at night. There is a light-dependent regulation of the cycle enzymes, as the third step requires reduced NADP.

There are two regulation systems at work when the cycle must be turned on or off: the thioredoxin/ferredoxin activation system, which activates some of the cycle enzymes; and the RuBisCo enzyme activation, active in the Calvin cycle, which involves its own activase.

The thioredoxin/ferredoxin system activates the enzymes glyceraldehyde-3-P dehydrogenase, glyceraldehyde-3-P phosphatase, fructose-1,6-bisphosphatase, sedoheptulose-1,7-bisphosphatase, and ribulose-5-phosphatase kinase, which are key points of the process. This happens when light is available, as the ferredoxin protein is reduced in the photosystem I complex of the thylakoid electron chain when electrons are circulating through it. Ferredoxin then binds to and reduces the thioredoxin protein, which activates the cycle enzymes by severing a cystine bond found in all these enzymes. This is a dynamic process as the same bond is formed again by other proteins that deactivate the enzymes. The implications of this process are that the enzymes remain mostly activated by day and are deactivated in the dark when there is no more reduced ferredoxin available.

The enzyme RuBisCo has its own, more complex activation process. It requires that a specific lysine amino acid be carbamylated to activate the enzyme. This lysine binds to RuBP and leads to a non-functional state if left uncarbamylated. A specific activase enzyme, called RuBisCo activase, helps this carbamylation process by removing one proton from the lysine and making the binding of the carbon dioxide molecule possible. Even then the RuBisCo enzyme is not yet functional, as it needs a magnesium ion bound to the lysine to function. This magnesium ion is released from the thylakoid lumen when the inner pH drops due to the active pumping of protons from the electron flow. RuBisCo activase itself is activated by increased concentrations of ATP in the stroma caused by its phosphorylation.

Chlorophyll

Chlorophyll is a green photosynthetic pigment found in plants, algae, and cyanobacteria. Chlorophyll is an essential component of photosynthesis, which helps plants get energy from light.

Chlorophyll absorbs most strongly in the blue and to a lesser extent red portions of the electromagnetic spectrum. The green portions of the electromagnetic spectrum, including wavelengths between five hundred and approximately six hundred nanometers, are not absorbed well, but are reflected by the plants, hence the green color of chlorophyll-containing tissues like plant leaves.

Chlorophyll, via its central role in photosynthesis, reflects harmony on both the sub-cellular and macro levels. On the sub-cellular level, the conversion of light energy via chlorphyll into useable chemical energy requies the complex coordination of several parts. On the higher level, the harmony between green plants, animals, and the environment is seen in the fact that plants use chlorophyll and water to convert carbon dioxide into glucose and free oxygen, while animals correspondingly use the oxygen and trapped energy stored in plant biomass and return carbon dioxide.

Chlorophyll's name is derived from ancient Greek: chloros = green and phyllon = leaf.

Chlorophyll gives leaves their green color.

Chlorophyll and Photosynthesis

In plant photosynthesis, incoming light is absorbed by chlorophyll and other accessory pigments in the antenna complexes of photosystem I and photosystem II. The antenna pigments are predominantly chlorophyll a, chlorophyll b, absorbing violet-blue and red light, respectively, and carotenoids. Their absorption spectra are non-overlapping, which serves to broaden the specific bandwidths of light these individual compounds absorb during the process of photosynthesis. The carotenoids also play a role as antioxidants, and serve to reduce photo-oxidative damage to chlorophyll molecules.

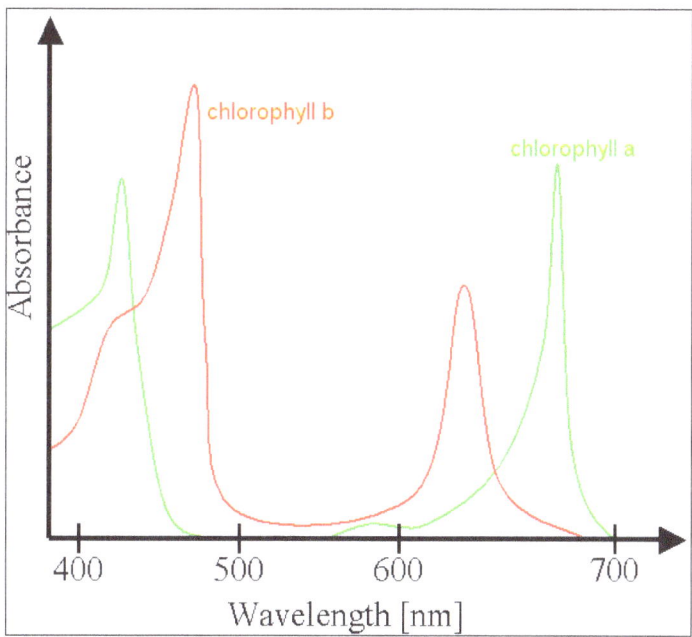

Absorbance spectra of free chlorophyll a (green) and b (red) in a solvent. The spectra of chlorophyll molecules are slightly modified in vivo depending on specific pigment-protein interactions.

Each antenna complex has between 250 and 400 pigment molecules, and the energy they absorb is shuttled by resonance energy transfer to a specialized chlorophyll a at the reaction center of each photosystem. When either of the two chorophyll a molecules at the reaction center absorb energy, an electron is excited and transferred to an electron-acceptor molecule, leaving an electron hole in the donor chlorophyll. In a poorly-understood reaction, electrons from water molecules participate in an oxidation reaction, where the hole from the donor chlorophyll is filled (recombined with another electron), and diatomic oxygen (O_2) is produced. Resulting chemical energy originating from the initial excited electron is eventually captured in the form of ATP and NADPH, and is then ultimately used to convert carbon dioxide (CO_2) to carbohydrates. This CO_2 fixation process results in the conversion (or an integrated external quantum efficiency) of 3 to 6 percent of the total incident solar radiation, with a theoretical maximum efficiency of 11 percent.

In Photosystem II, the electron which reduces P680+ ultimately comes from the oxidation of water into O_2 and H^+ through several intermediates. This reaction is how photosynthetic organisms like plants produce O_2 gas, and is the source for practically all the O_2 in Earth's atmosphere.

Special Pair

The photosystem reaction centers consist of a "special pair" of chlorophyll a molecules that are characterized by their specific absorption maximum. The special pair in photosystem I are designated P700, and those from photosystem II are designated P680. The P is short for pigment, and the number is the specific absorption peak in nanometers for the chlorophyll molecules in each reaction center.

Chlorophyll a is common to all eukaryotic photosynthetic organisms, and, due to its central role in the reaction center, is essential for photosynthesis. The accessory pigments such as chlorophyll b and carotenoids are not essential. Some algae, such as brown algae and diatoms, use chlorophyll c as a substitute for chlorophyll b. Historically, red algae have been assumed to have chlorophyll d, although it could not be isolated from all species and even different collections of the same species. This puzzle has recently been resolved, since the chlorophyll d is actually from an epiphytic cyanobacterium (Acaryochloris marina) that lives on the red algae. These cyanobacteria have a ratio of chlorophyll d: chlorophyll a of approximately 30 to 1, and represent a rare example of a photosystem with chlorophyll d at the reaction center of the photosystem. All other known eukaryotes and cyanobacteria use chlorophyll a. There are likely to be many chlorophyll-d containing organisms awaiting discovery, for example a free living form was recently found in the Salton Sea.

Other chemical variations of chlorophyll are found in photosynthetic bacteria, other than cyanobacteria (sometimes called blue-green algae). Purple bacteria use bacteriochlorophyll, which absorbs infrared light between 800 to 1000 nanometers, and the green sulphur bacteria use chlorobium chlorophyll. All known bacteria with bacteriochlorophyll have a form of photosynthesis that does not involve evolution of oxygen and so are called anoxyphotobacteria. There is a very large number of different bacteriochlorophylls in different anoxyphotobacteria, including one species which contains zinc, rather than the usual magnesium as the coordinated metal.

Because the different chlorophyll and non-chlorophyll pigments associated with the photosystems all have different spectra, the total absorption spectrum is broadened and flattened such that a wider range of red, orange, yellow, and blue light can be absorbed by plants and algae. Most photosynthetic organisms do not have pigments which absorb green light well, thus most remaining light under leaf canopies in forests or under water with abundant plankton is green, a spectral effect called the "green window." Some organisms, such as cyanobacteria and red algae, contain accessory phycobilin pigments that can absorb green light relatively well and thus they can exploit the little remaining green light in these habitats.

Chemical Structure

	Chlorophyll a	Chlorophyll b	Chlorophyll c1	Chlorophyll c2	Chlorophyll d
Molecular formula	$C_{55}H_{72}O_5N_4Mg$	$C_{55}H_{70}O_6N_4Mg$	$C_{35}H_{30}O_5N_4Mg$	$C_{35}H_{28}O_5N_4Mg$	$C_{54}H_{70}O_6N_4Mg$
C3 group	-CH=CH$_2$	-CH=CH$_2$	-CH=CH$_2$	-CH=CH$_2$	-CHO
C7 group	-CH$_3$	-CHO	-CH$_3$	-CH$_3$	-CH$_3$
C8 group	-CH$_2$CH$_3$	-CH$_2$CH$_3$	-CH$_2$CH$_3$	-CH=CH$_2$	-CH$_2$CH$_3$
C17 group	-CH$_2$CH-$_2$COO-Phytyl	-CH$_2$CH-$_2$COO-Phytyl	-CH=CHCOOH	-CH=CHCOOH	-CH$_2$CH-$_2$COO-Phytyl
C17-C18 bond	Single	Single	Double	Double	Single
Occurrence	Universal	Mostly plants	Various algae	Various algae	Cyanobacteria

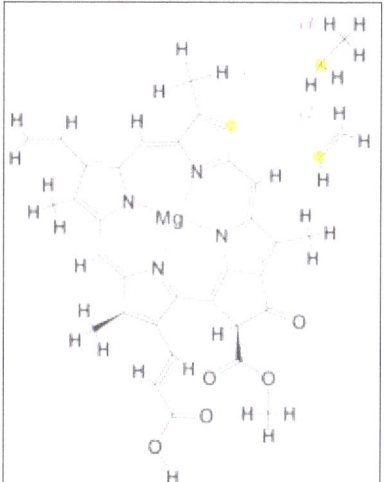

Common structure of chlorophyll a, b and d Common structure of chlorophyll c1, and c2.

Chlorophyll can be shown to be vital for photosynthesis by destarching a leaf from a variegated plant and exposing it to light for several hours (variegated leaves have green areas that contain chlorophyll and white areas that have none). When tested with iodine solution, a color change revealing the presence of starch occurs only in regions of the leaf that were green and therefore contained chlorophyll. This shows that photosynthesis does not occur in areas where chlorophyll is absent, and constitutes evidence that the presence of chlorophyll is a requirement for photosynthesis.

Photosynthetic Efficiency

The photosynthetic efficiency is the fraction of light energy converted into chemical energy during photosynthesis in plants and algae. Photosynthesis can be described by the simplified chemical reaction.

$$6H_2O + 6CO_2 + energy \rightarrow C_6H_{12}O_6 + 6O_2$$

where $C_6H_{12}O_6$ is glucose (which is subsequently transformed into other sugars, cellulose, lignin, and so forth). The value of the photosynthetic efficiency is dependent on how light energy is defined – it depends on whether we count only the light that is absorbed, and on what kind of light is used. It takes eight (or perhaps 10 or more) photons to utilize one molecule of CO_2. The Gibbs free energy for converting a mole of CO_2 to glucose is 114 kcal, whereas eight moles of photons of wavelength 600 nm contains 381 kcal, giving a nominal efficiency of 30%. However, photosynthesis can occur with light up to wavelength 720 nm so long as there is also light at wavelengths below 680 nm to keep Photosystem II operating. Using longer wavelengths means less light energy is needed for the same number of photons and therefore for the same amount of photosynthesis. For actual sunlight, where only 45% of the light is in the photosynthetically active wavelength range, the theoretical maximum efficiency of solar energy conversion is approximately 11%. In actuality, however, plants do not absorb all incoming sunlight (due to reflection, respiration requirements of photosynthesis and the need for optimal solar radiation levels) and do not convert all harvested energy into biomass, which results in a maximum overall photosynthetic efficiency of 3 to 6% of total solar radiation. If photosynthesis is inefficient, excess light energy must be dissipated to avoid damaging the photosynthetic apparatus. Energy can be dissipated as heat (non-photochemical quenching), or emitted as chlorophyll fluorescence.

Typical Efficiencies

Plants

Quoted values sunlight-to-biomass efficiency

Plant	Efficiency
Plants, typical	0.1% 0.2–2%3.5-4.3%
Typical crop plants	1–2%

The following is a breakdown of the energetics of the photosynthesis process from Photosynthesis by Hall and Rao:

Starting with the solar spectrum falling on a leaf,

47% lost due to photons outside the 400–700 nm active range (chlorophyll utilizes photons, between 400 and 700 nm, extracting the energy of one 700 nm photon from each one) 30% of the in-band photons are lost due to incomplete absorption or photons hitting components other than chloroplasts. 24% of the absorbed photon energy is lost due to degrading short wavelength photons to the 700 nm energy level, 68% of the utilized energy is lost in conversion into d-glucose, 35–45% of the glucose is consumed by the leaf in the processes of dark and photo respiration.

Stated another way:

100% sunlight → non-bioavailable photons waste is 47%, leaving

53% (in the 400–700 nm range) → 30% of photons are lost due to incomplete absorption, leaving 37% (absorbed photon energy) → 24% is lost due to wavelength-mismatch degradation to 700 nm energy, leaving.

28.2% (sunlight energy collected by chlorophyll) → 32% efficient conversion of ATP and NADPH to d-glucose, leaving.

9% (collected as sugar) → 35–40% of sugar is recycled/consumed by the leaf in dark and photo-respiration, leaving.

5.4% net leaf efficiency.

Many plants lose much of the remaining energy on growing roots. Most crop plants store ~0.25% to 0.5% of the sunlight in the product (corn kernels, potato starch, etc.).

Photosynthesis increases linearly with light intensity at low intensity, but at higher intensity this is no longer the case. Above about 10,000 lux or ~100 watts/square meter the rate no longer increases. Thus, most plants can only utilize ~10% of full mid-day sunlight intensity. This dramatically reduces average achieved photosynthetic efficiency in fields compared to peak laboratory results. However, real plants (as opposed to laboratory test samples) have lots of redundant, randomly oriented leaves. This helps to keep the average illumination of each leaf well below the mid-day peak enabling the plant to achieve a result closer to the expected laboratory test results using limited illumination.

Only if the light intensity is above a plant specific value, called the compensation point the plant assimilates more carbon and releases more oxygen by photosynthesis than it consumes by cellular respiration for its own current energy demand.

Photosynthesis measurement systems are not designed to directly measure the amount of light absorbed by the leaf. Nevertheless, the light response curves that the class produces do allow comparisons in photosynthetic efficiency between plants.

Algae and other Monocellular Organisms

From a 2010 study by the University of Maryland, photosynthesizing Cyanobacteria have been shown to be a significant species in the global carbon cycle, accounting for 20–30% of Earth's photosynthetic productivity and convert solar energy into biomass-stored chemical energy at the rate of ~450 TW. The efficiency of algae is much higher compared to that of plants (98 percent efficiency for algae compared to just 12 percent in plants).

Efficiencies of Various Biofuel Crops

Popular choices for plant biofuels include: oil palm, soybean, castor oil, sunflower oil, safflower oil, corn ethanol, and sugar cane ethanol.

An analysis of a proposed Hawaiian oil palm plantation claimed to yield 600 gallons of biodiesel per acre per year. That comes to 2835 watts per acre or 0.7 W/m². Typical insolation in Hawaii is around 5.5 kWh/(m²day) or 230 W/m². For this particular oil palm plantation, if it delivered the claimed 600 gallons of biodiesel per acre per year, would be converting 0.3% of the incident solar energy to chemical fuel. Total photosynthetic efficiency would include more than just the biodiesel oil, so this 0.3% number is something of a lower bound.

Contrast this with a typical photovoltaic installation, which would produce an average of roughly 22 W/m² (roughly 10% of the average insolation), throughout the year. Furthermore, the photovoltaic panels would produce electricity, which is a high-quality form of energy, whereas converting the biodiesel into mechanical energy entails the loss of a large portion of the energy. On the other hand, a liquid fuel is much more convenient for a vehicle than electricity, which has to be stored in heavy, expensive batteries.

Most crop plants store ~0.25% to 0.5% of the sunlight in the product (corn kernels, potato starch, etc.), sugar cane is exceptional in several ways to yield peak storage efficiencies of ~8%.

Ethanol fuel in Brazil has a calculation that results in: "Per hectare per year, the biomass produced corresponds to 0.27 TJ. This is equivalent to 0.86 W/m². Assuming an average insolation of 225 W/m², the photosynthetic efficiency of sugar cane is 0.38%." Sucrose accounts for little more than 30% of the chemical energy stored in the mature plant; 35% is in the leaves and stem tips, which are left in the fields during harvest, and 35% are in the fibrous material (bagasse) left over from pressing.

C3 vs. C4 and CAM plants

C3 plants use the Calvin cycle to fix carbon. C4 plants use a modified Calvin cycle in which they separate Ribulose-1,5-bisphosphate carboxylase oxygenase (RuBisCO) from atmospheric oxygen, fixing carbon in their mesophyll cells and using oxaloacetate and malate to ferry the fixed carbon to RuBisCO and the rest of the Calvin cycle enzymes isolated in the bundle-sheath cells. The intermediate compounds both contain four carbon atoms, which gives C4. In Crassulacean acid metabolism (CAM), time isolates functioning RuBisCo (and the other Calvin cycle enzymes) from high oxygen concentrations produced by photosynthesis, in that O_2 is evolved during the day, and allowed to dissipate then, while at night atmospheric CO_2 is taken up and stored as malic or other acids. During the day, CAM plants close stomata and use stored acids as carbon sources for sugar, etc. production.

The C3 pathway requires 18 ATP and 12 NADPH for the synthesis of one molecule of glucose (3 ATP + 2 NADPH per CO_2 fixed) while the C4 pathway requires 30 ATP and 12 NADPH (C3 + 2 ATP per CO2 fixed). In addition, we can take into account that each NADPH is equivalent to 3 ATP, that means both pathways require 36 additional (equivalent of) ATP. Despite this reduced ATP efficiency, C4 is an evolutionary advancement, adapted to areas of high levels of light, where the reduced ATP efficiency is more than offset by the use of increased light. The ability to thrive despite restricted water availability maximizes the ability to use available light. The simpler C3 cycle which operates in most plants is adapted to wetter darker environments, such as many northern latitudes corn, sugar cane, and sorghum are C4 plants. These plants are economically important in part because of their relatively high photosynthetic efficiencies compared to many other crops. Pineapple is a CAM plant.

Anoxygenic Photosynthesis

Bacterial anoxygenic photosynthesis is distinguished from the more familiar terrestrial plant oxygenic photosynthesis by the nature of the terminal reductant (e.g. hydrogen sulfide rather than water) and in the byproduct generated (e.g. elemental sulfur instead of molecular oxygen). As its name implies, anoxygenic photosynthesis does not produce oxygen as a byproduct of the reaction. Several groups of bacteria can conduct anoxygenic photosynthesis: green sulfur bacteria (GSB), red and green filamentous phototrophs (FAPs e.g. Chloroflexi), purple bacteria, Acidobacteria, and heliobacteria.

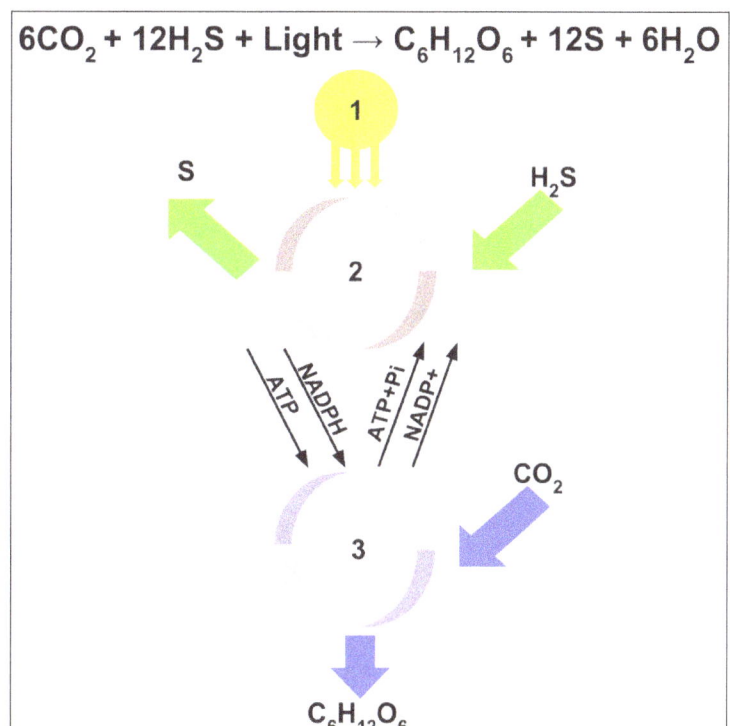

$$6CO_2 + 12H_2S + Light \rightarrow C_6H_{12}O_6 + 12S + 6H_2O$$

Sulfur is used as a reducing agent during photosynthesis in green and sulfur bacteria. 1. Energy in the form of sunlight. 2. The light dependent reactions take place when the light excites a reaction center, which donates an electron to another molecule and starts the electron transport chain to produce ATP and NADPH. 3. Once NADPH has been produced, the Calvin cycle proceeds as in oxygenic photosynthesis, turning CO2 into glucose.

The pigments used to carry out anaerobic photosynthesis are similar to chlorophyll but differ in molecular detail and peak wavelength of light absorbed. Bacteriochlorophylls a through g absorb electromagnetic photons maximally in the near-infrared within their natural membrane milieu. This differs from chlorophyll a, the predominant plant and cyanobacteria pigment, which has peak absorption wavelength approximately 100 nanometers shorter (in the red portion of the visible spectrum).

Some archaea (e.g. Halobacterium) capture light energy for metabolic function and are thus phototrophic but none are known to "fix" carbon (i.e. be photosynthetic). Instead of a chlorophyll-type receptor and electron transport chain, proteins such as halorhodopsin capture light energy with the aid of diterpenes to move ions against the gradient and produce ATP via chemiosmosis in the manner of mitochondria.

There are two main types of anaerobic photosynthetic electron transport chains in bacteria. The type I reaction centers found in GSB, Chloracidobacterium, and Heliobacteria and the type II reaction centers found in FAPs and Purple Bacteria.

Type I Reaction Centers

The electron transport chain of green sulfur bacteria — such as is present in model organism Chlorobaculum tepidum — uses the reaction centre bacteriochlorophyll pair, P840. When light is absorbed by the reaction center, P840 enters an excited state with a large negative reduction potential, and so readily donates the electron to bacteriochlorophyll 663 which passes it on down the electron chain. The electron is transferred through a series of electron carriers and complexes until it is used to reduce NAD^+. P840 regeneration is accomplished with the oxidation of sulfide ion from hydrogen sulfide (or hydrogen or ferrous iron) by cytochrome c_{555}.

Type II Reaction Centers

Although the type II reaction centers are structurally and sequentially analogous to Photosystem II (PSII) in plant chloroplasts and cyanobacteria, known organisms that exhibit anoxygenic photosynthesis do not have a region analogous to the oxygen-evolving complex of PSII.

The electron transport chain of purple non-sulfur bacteria begins when the reaction centre bacteriochlorophyll pair, P870, becomes excited from the absorption of light. Excited P870 will then donate an electron to bacteriopheophytin, which then passes it on to a series of electron carriers down the electron chain. In the process, it will generate an electro-chemical gradient which can then be used to synthesize ATP by chemiosmosis. P870 has to be regenerated (reduced) to be available again for a photon reaching the reaction-center to start the process anew. Molecular hydrogen in the bacterial environment is the usual electron donor.

Factors Affecting Photosynthesis

Light

It is one of the major factors affecting photosynthesis. Photosynthesis cannot occur in the dark and the source of light for the plants is sunlight. Three attributes of light are important for photosynthesis:

- Intensity: Photosynthesis begins at low intensities of light and increases till it is maximum at the brightest time of the day. The amount of light required varies for different plants. Photosynthesis uses maximum up to 1.5 % light in the process and so light is generally not a limiting factor at high intensity. However, the light becomes a limiting factor in low intensity because no matter how much water or CO_2 is present, without light photosynthesis cannot occur. At high intensities, the temperature of the plant increases which leads to increased transpiration in the plant. This leads to the closing of the stomata which leads to a reduced CO2 intake. Thus, leading to a reduction and finally stoppage of photosynthesis. Therefore, excessive light inhibits photosynthesis.

- Quality: Experiments conducted by Engelmann prove that the chlorophyll most effectively absorbs red and blue wavelengths from the entire spectrum of light. Thus, maximum photosynthesis occurs when the plant is exposed to the light of these wavelengths.

- Duration: The longer the plant is exposed to light, the longer the process of photosynthesis will continue. As long as the temperature of the plant remains balanced, photosynthesis will occur.

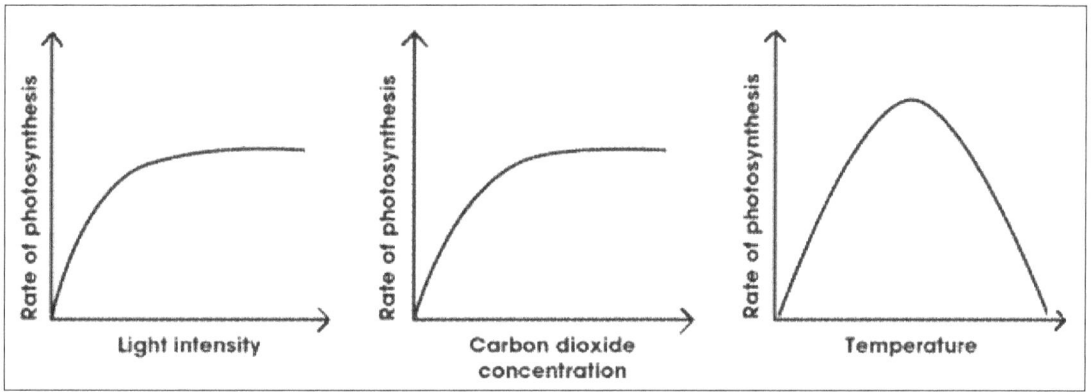

Carbon Dioxide Concentration

The atmosphere contains 0.03% of carbon dioxide amidst other gases. Plants take in carbon dioxide from the air. But, since the amount of CO_2 in the air is very less, it acts as a limiting factor for photosynthesis. Experiments have been performed to study the rate of photosynthesis on increasing the concentration of CO_2 in the atmosphere.

It is seen that, when light and temperature are not the limiting factors, increasing CO2 concentration leads to an increase in the rate of photosynthesis. But, beyond a certain limit, CO_2 starts accumulating in the plant and this leads to slowing down of the process. So, excessive CO_2 inhibits photosynthesis especially when it starts to accumulate.

Temperature

It is commonly seen in all biological and biochemical processes that they occur best in a certain optimum range of temperature. This holds true for photosynthesis as well. It is observed that, when CO_2 and light are not limiting factors, the rate of photosynthesis increases with increase in temperatures till the optimum level for that plant. Beyond the optimum levels on both sides of the normal range, the enzymes are deactivated or destroyed and photosynthesis stops.

Water

Water is considered one of the most important factors affecting photosynthesis. When there is a reduced water intake or availability, the stomata begin to close to avoid loss of any water during transpiration. With the stomata closing down the CO2 intake also stops which affects photosynthesis. Therefore, the effect of water on photosynthesis is more indirect than direct.

Oxygen

Optimum levels of oxygen are favourable for photosynthesis. Oxygen is needed for photorespiration in C3 plants and the by-product of photorespiration is CO_2 which is essential for photosynthesis. Also, the energy generated during the oxygen respiration is needed for the process of photosynthesis as well. However, an increase in the oxygen levels beyond the optimum for the plant leads to inhibition of photosynthesis.

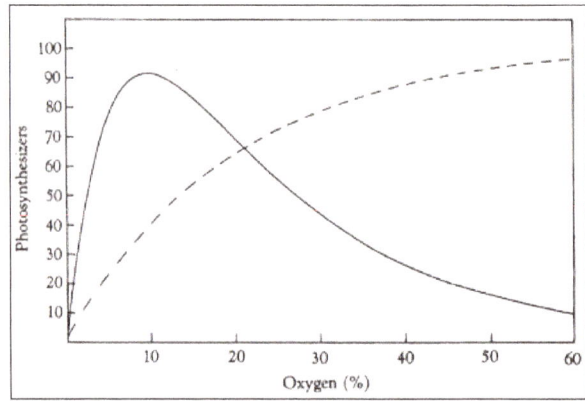

This is because oxygen tends to break down the intermediaries that are formed in photosynthesis. Oxygen also completes with CO2 to combine with RUBISCO which a part of the dark reaction of photosynthesis and photorespiration. Therefore, increased levels of O2 would mean that RUBISCO will combine with O2 to initiate photorespiration and photosynthesis will slow down.

Blackman's Principle of Limiting Factors

This principle states that when a process is governed by more than one factor, the rate of the process is governed by that factor which is closest to its minimum value.

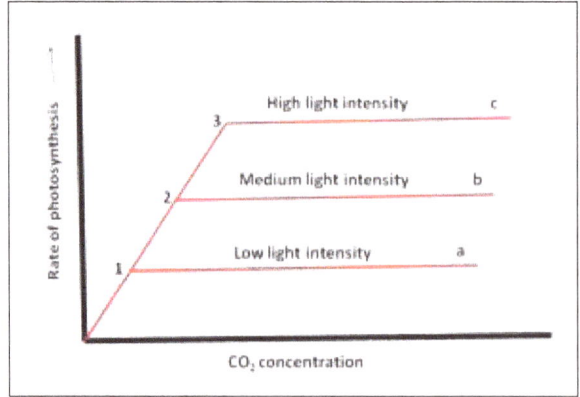

For example, if a leaf is exposed to a certain amount of light intensity with constant temperature but the CO_2 available is less, the rate of photosynthesis will not increase with an increase in the light intensity. Therefore, in this case, CO_2 is the limiting factor.

Application of Photosynthesis

Agriculture

The process of natural photosynthesis is a highly inefficient process. For example, the typical efficiency of plants is 0.1% or 0.2-2%, the typical efficiency of crop plants is 1-2%, and the plant which has the maximum efficiency of 7-8% is sugarcane. The inefficiency of plants to convert sunlight into viable energy sources for humans has prompted agriculturalists to attempt to enhance the abilities of plant to use more of the energy that they absorb. In order to more fully understand how to control the photosynthetic efficiency of plants, agriculturalists need to understand the process of photo respiration, or the light-independent reactions of photosynthesis. These are the photosynthetic reactions that use carbon dioxide and convert it into sugars for the plants to use for growth.

On a global basis, photosynthesis plays a large role in politics and the global food market. As the images depict above the United Sate and China are two of the largest contributors, financially, to agriculture production across the world, however the U.S. and Canada also consume significantly more food on a daily basis than any other country. Persons residing in the U.S. and Canada consume over 3,400 calories of food on average per day, while persons in central Africa consume less than 1,800 calories on average per day. These averages do not reflect the personal consumption of yourself or your family, however from these images we can learn how the U.S. compares to other countries in the world.

Energy

The majority of substances that we use to fuel our everyday energy needs come from the breakdown or burning of plants. We use fossils fuels, which are formed by the decomposition of plant materials over millions of years, we burn wood and coal to heat houses and warm fireplaces, and in recent years we have produced a product called Ethanol, which comes from the sugars and starched in corn and can be used to fuel cars, buses, and trucks. Ultimately all of the energy we use today comes from the process of photosynthesis which captured the energy of the sun and stored it in carbohydrates such as sugar and starches.

Materials

Many of the materials that are used in our daily lives depend, either directly or indirectly, on photosynthesis. The paper in notebook and books comes directly from photosynthesis. Paper is made mostly of cellulose, a compound formed by stringing many sugar molecules together to form a starch. These sugars come directly from the process of photosynthesis and cellulose is what gives a plant its structure. An example of a material that is indirectly linked to photosynthesis is plastic. Plastics are made from petroleum products which is the basic fossil fuel we use to make gas for cars, motor oil and many other products.

The Environment

When it comes to photosynthesis, one of the major environmental concerns associated with it is the consumption and production of carbon dioxide gas (CO_2). Photosynthesis requires carbon dioxide gas in order to produce the sugars plants need to grow and survive. In this sense photosynthesis helps balance the amount of carbon dioxide that is present in our atmosphere. Naturally, photosynthesis maintains a steady balance of carbon dioxide in the atmosphere. However, when plant material is consumed, burned, by humans for fuel the carbon dioxide that was stored in the plant material is released in quantities that are too high for photosynthesis to balance naturally. The burning of materials such as fossil fuels and wood cause a buildup of carbon dioxide gas in the Earth's atmosphere. As you can see in the image shown below, the carbon dioxide in the atmosphere traps heat from the sun and causes the Earth to warm. Over the past few decades, scientists have discovered that the build-up of carbon dioxide in the atmosphere is contributing to the warming of the Earth and climate change.

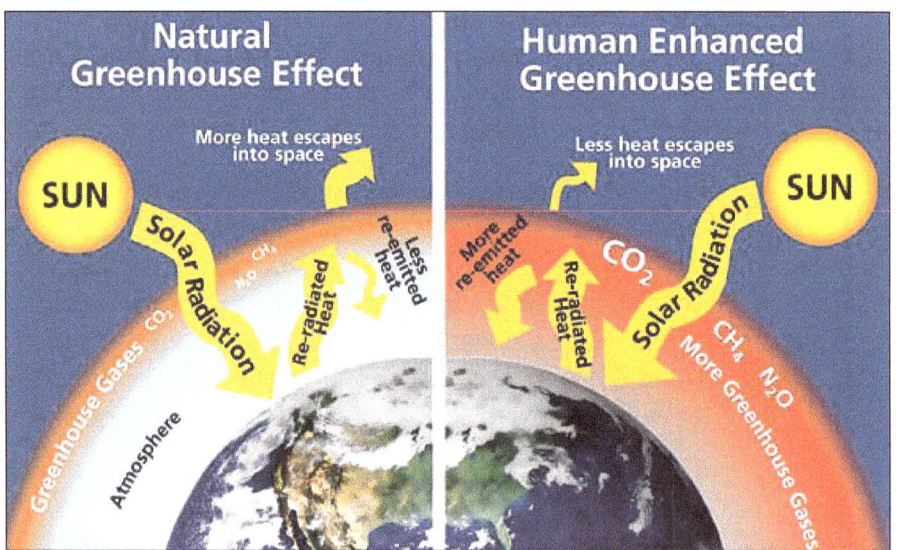

Artificial Photosynthesis

Artificial photosynthesis is a chemical process that biomimics the natural process of photosynthesis to convert sunlight, water, and carbon dioxide into carbohydrates and oxygen. The term artificial photosynthesis is commonly used to refer to any scheme for capturing and storing the energy from sunlight in the chemical bonds of a fuel (a solar fuel). Photocatalytic water splitting converts water into hydrogen and oxygen and is a major research topic of artificial photosynthesis. Light-driven carbon dioxide reduction is another process studied that replicates natural carbon fixation.

Research of this topic includes the design and assembly of devices for the direct production of solar fuels, photoelectrochemistry and its application in fuel cells, and the engineering of enzymes and photoautotrophic microorganisms for microbial biofuel and biohydrogen production from sunlight.

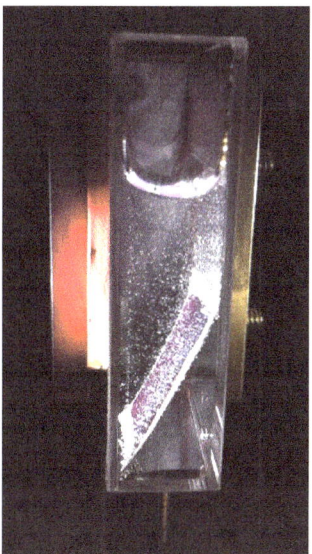

A sample of a photoelectric cell in a lab environment.

Catalysts are added to the cell, which is submerged in water and illuminated by simulated sunlight. The bubbles seen are oxygen (forming on the front of the cell) and hydrogen (forming on the back of the cell).

The photosynthetic reaction can be divided into two half-reactions of oxidation and reduction, both of which are essential to producing fuel. In plant photosynthesis, water molecules are photo-oxidized to release oxygen and protons. The second phase of plant photosynthesis (also known as the Calvin-Benson cycle) is a light-independent reaction that converts carbon dioxide into glucose (fuel). Researchers of artificial photosynthesis are developing photocatalysts that are able to perform both of these reactions. Furthermore, the protons resulting from water splitting can be used for hydrogen production. These catalysts must be able to react quickly and absorb a large percentage of the incident solar photons.

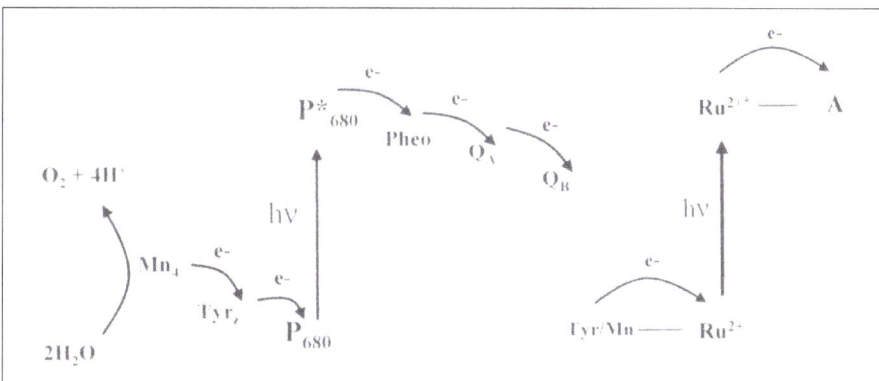

Natural (left) versus artificial photosynthesis (right).

Whereas photovoltaics can provide energy directly from sunlight, the inefficiency of fuel production from photovoltaic electricity (indirect process) and the fact that sunshine is not constant throughout the day sets a limit to its use. One way of using natural photosynthesis is for the production of a biofuel, which is an indirect process that suffers from low energy conversion efficiency (due to photosynthesis' own low efficiency in converting sunlight to biomass), the cost of harvesting and transporting the fuel, and conflicts due to the increasing need of land mass for food production. The purpose of artificial photosynthesis is to produce a fuel from sunlight that can be stored conveniently and used when sunlight is not available, by using direct processes, that is, to produce a solar fuel. With the development of catalysts able to reproduce the major parts of photosynthesis, water and sunlight would ultimately be the only needed sources for clean energy production. The only by-product would be oxygen, and production of a solar fuel has the potential to be cheaper than gasoline.

One process for the creation of a clean and affordable energy supply is the development of photocatalytic water splitting under solar light. This method of sustainable hydrogen production is a major objective for the development of alternative energy systems. It is also predicted to be one of the more, if not the most, efficient ways of obtaining hydrogen from water. The conversion of solar energy into hydrogen via a water-splitting process assisted by photosemiconductor catalysts is one of the most promising technologies in development. This process has the potential for large quantities of hydrogen to be generated in an ecologically sound manner. The conversion of solar energy into a clean fuel (H_2) under ambient conditions is one of the greatest challenges facing scientists in the twenty-first century.

Two methods are generally recognized for the construction of solar fuel cells for hydrogen production:

- A homogeneous system is one such that catalysts are not compartmentalized, that is, components are present in the same compartment. This means that hydrogen and oxygen are produced in the same location. This can be a drawback, since they compose an explosive mixture, demanding gas product separation. Also, all components must be active in approximately the same conditions (e.g., pH).

- A heterogeneous system has two separate electrodes, an anode and a cathode, making possible the separation of oxygen and hydrogen production. Furthermore, different components do not necessarily need to work in the same conditions. However, the increased complexity of these systems makes them harder to develop and more expensive.

Another area of research within artificial photosynthesis is the selection and manipulation of photosynthetic microorganisms, namely green microalgae and cyanobacteria, for the production of solar fuels. Many strains are able to produce hydrogen naturally, and scientists are working to improve them. Algae biofuels such as butanol and methanol are produced both at laboratory and commercial scales. This method has benefited from the development of synthetic biology, which is also being explored by the J. Craig Venter Institute to produce a synthetic organism capable of biofuel production. In 2017, an efficient process was developed to produce acetic acid from carbon dioxide using "cyborg bacteria".

In energy terms, natural photosynthesis can be divided in three steps:

- Light-harvesting complexes in bacteria and plants capture photons and transduce them into electrons, injecting them into the photosynthetic chain.

- Proton-coupled electron transfer along several cofactors of the photosynthetic chain, causing local, spatial charge separation.

- Redox catalysis, which uses the aforementioned transferred electrons to oxidize water to dioxygen and protons; these protons can in some species be utilized for dihydrogen production.

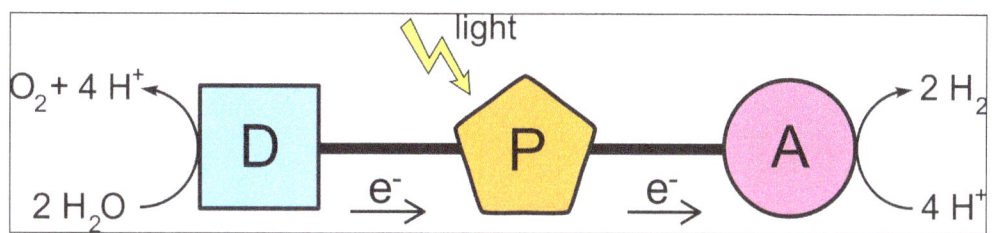

A triad assembly, with a photosensitizer (P) linked in tandem to a water oxidation catalyst (D) and a hydrogen evolving catalyst (A). Electrons flow from D to A when catalysis occurs.

Using biomimetic approaches, artificial photosynthesis tries to construct systems doing the same type of processes. Ideally, a triad assembly could oxidize water with one catalyst, reduce protons with another and have a photosensitizer molecule to power the whole system. One of the simplest designs is where the photosensitizer is linked in tandem between a water oxidation catalyst and a hydrogen evolving catalyst:

- The photosensitizer transfers electrons to the hydrogen catalyst when hit by light, becoming oxidized in the process.

- This drives the water splitting catalyst to donate electrons to the photosensitizer. In a triad assembly, such a catalyst is often referred to as a donor. The oxidized donor is able to perform water oxidation.

The state of the triad with one catalyst oxidized on one end and the second one reduced on the other end of the triad is referred to as a charge separation, and is a driving force for further electron transfer, and consequently catalysis, to occur. The different components may be assembled in diverse ways, such as supramolecular complexes, compartmentalized cells, or linearly, covalently linked molecules.

Research into finding catalysts that can convert water, carbon dioxide, and sunlight to carbohydrates or hydrogen is a current, active field. By studying the natural oxygen-evolving complex (OEC), researchers have developed catalysts such as the "blue dimer" to mimic its function or inorganic-based materials such as Birnessite with the similar building block as the OEC. Photoelectrochemical cells that reduce carbon dioxide into carbon monoxide (CO), formic acid (HCOOH) and methanol (CH_3OH) are under development. However, these catalysts are still very inefficient.

Hydrogen Catalysts

Hydrogen is the simplest solar fuel to synthesize, since it involves only the transference of two electrons to two protons. It must, however, be done stepwise, with formation of an intermediate hydride anion:

$$2\,e^- + 2\,H^+ \rightleftharpoons H^+ + H^- \rightleftharpoons H_2$$

The proton-to-hydrogen converting catalysts present in nature are hydrogenases. These are enzymes that can either reduce protons to molecular hydrogen or oxidize hydrogen to protons and electrons. Spectroscopic and crystallographic studies spanning several decades have resulted in a good understanding of both the structure and mechanism of hydrogenase catalysis. Using this information, several molecules mimicking the structure of the active site of both nickel-iron and iron-iron hydrogenases have been synthesized. Other catalysts are not structural mimics of hydrogenase but rather functional ones. Synthesized catalysts include structural H-cluster models, a dirhodium photocatalyst, and cobalt catalysts.

Water-Oxidizing Catalysts

Water oxidation is a more complex chemical reaction than proton reduction. In nature, the oxygen-evolving complex performs this reaction by accumulating reducing equivalents (electrons) in a manganese-calcium cluster within photosystem II (PS II), then delivering them to water molecules, with the resulting production of molecular oxygen and protons:

$$2\ H_2O \rightarrow O_2 + 4\ H^+ + 4e^-$$

Without a catalyst (natural or artificial), this reaction is very endothermic, requiring high temperatures (at least 2500 K).

The exact structure of the oxygen-evolving complex has been hard to determine experimentally. As of 2011, the most detailed model was from a 1.9 Å resolution crystal structure of photosystem II. The complex is a cluster containing four manganese and one calcium ions, but the exact location and mechanism of water oxidation within the cluster is unknown. Nevertheless, bio-inspired manganese and manganese-calcium complexes have been synthesized, such as $[Mn_4O_4]$ cubane-type clusters, some with catalytic activity.

Some ruthenium complexes, such as the dinuclear μ-oxo-bridged "blue dimer" (the first of its kind to be synthesized), are capable of light-driven water oxidation, thanks to being able to form high valence states. In this case, the ruthenium complex acts as both photosensitizer and catalyst.

Many metal oxides have been found to have water oxidation catalytic activity, including ruthenium(IV) oxide (RuO_2), iridium(IV) oxide (IrO_2), cobalt oxides (including nickel-doped Co_3O_4), manganese oxide (including layered MnO_2 (birnessite), Mn_2O_3), and a mix of Mn_2O_3 with $CaMn_2O_4$. Oxides are easier to obtain than molecular catalysts, especially those from relatively abundant transition metals (cobalt and manganese), but suffer from low turnover frequency and slow electron transfer properties, and their mechanism of action is hard to decipher and, therefore, to adjust.

Recently Metal-Organic Framework (MOF)-based materials have been shown to be a highly promising candidate for water oxidation with first row transition metals. The stability and tunability of this system is projected to be highly beneficial for future development.

Photosensitizers

Nature uses pigments, mainly chlorophylls, to absorb a broad part of the visible spectrum. Artificial systems can use either one type of pigment with a broad absorption range or combine several pigments for the same purpose.

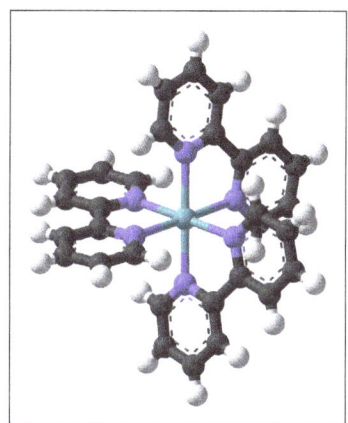

Structure of [Ru(bipy)$_3$]$^{2+}$, a broadly used photosensitizer.

Ruthenium polypyridine complexes, in particular tris(bipyridine)ruthenium(II) and its derivatives, have been extensively used in hydrogen photoproduction due to their efficient visible light absorption and long-lived consequent metal-to-ligand charge transfer excited state, which makes the complexes strong reducing agents. Other noble metal-containing complexes used include ones with platinum, rhodium and iridium.

Metal-free organic complexes have also been successfully employed as photosensitizers. Examples include eosin Y and rose bengal. Pyrrole rings such as porphyrins have also been used in coating nanomaterials or semiconductors for both homogeneous and heterogeneous catalysis.

As part of current research efforts artificial photonic antenna systems are being studied to determine efficient and sustainable ways to collect light for artificial photosynthesis. Gion Calzaferri (2009) describes one such antenna that uses zeolite L as a host for organic dyes, to mimic plant's light collecting systems. The antenna is fabricated by inserting dye molecules into the channels of zeolite L. The insertion process, which takes place under vacuum and at high temperature conditions, is made possible by the cooperative vibrational motion of the zeolite framework and of the dye molecules. The resulting material may be interfaced to an external device via a stopcock intermediate.

Carbon Dioxide Reduction Catalysts

In nature, carbon fixation is done by green plants using the enzyme RuBisCO as a part of the Calvin cycle. RuBisCO is a rather slow catalyst compared to the vast majority of other enzymes, incorporating only a few molecules of carbon dioxide into ribulose-1,5-bisphosphate per minute, but does so at atmospheric pressure and in mild, biological conditions. The resulting product is further reduced and eventually used in the synthesis of glucose, which in turn is a precursor to more complex carbohydrates, such as cellulose and starch. The process consumes energy in the form of ATP and NADPH.

Artificial CO_2 reduction for fuel production aims mostly at producing reduced carbon compounds from atmospheric CO_2. Some transition metal polyphosphine complexes have been developed for this end; however, they usually require previous concentration of CO_2 before use, and carriers (molecules that would fixate CO_2) that are both stable in aerobic conditions and able to concentrate CO_2 at atmospheric concentrations haven't been yet developed. The simplest product from CO_2

reduction is carbon monoxide (CO), but for fuel development, further reduction is needed, and a key step also needing further development is the transfer of hydride anions to CO.

Other Materials and Components

Charge separation is a major property of dyad and triad assemblies. Some nanomaterials employed are fullerenes (such as carbon nanotubes), a strategy that explores the pi-bonding properties of these materials. Diverse modifications (covalent and non-covalent) of carbon nanotubes have been attempted to increase the efficiency of charge separation, including the addition of ferrocene and pyrrole-like molecules such as porphyrins and phthalocyanines.

Since photodamage is usually a consequence in many of the tested systems after a period of exposure to light, bio-inspired photoprotectants have been tested, such as carotenoids (which are used in photosynthesis as natural protectants).

Light-Driven Methodologies Under Development

Photoelectrochemical Cells

Photoelectrochemical cells are a heterogeneous system that use light to produce either electricity or hydrogen. The vast majority of photoelectrochemical cells use semiconductors as catalysts. There have been attempts to use synthetic manganese complex-impregnated Nafion as a working electrode, but it has been since shown that the catalytically active species is actually the broken-down complex.

A promising, emerging type of solar cell is the dye-sensitized solar cell. This type of cell still depends on a semiconductor (such as TiO_2) for current conduction on one electrode, but with a coating of an organic or inorganic dye that acts as a photosensitizer; the counter electrode is a platinum catalyst for H_2 production. These cells have a self-repair mechanism and solar-to-electricity conversion efficiencies rivaling those of solid-state semiconductor ones.

Photocatalytic Water Splitting in Homogeneous Systems

Direct water oxidation by photocatalysts is a more efficient usage of solar energy than photoelectrochemical water splitting because it avoids an intermediate thermal or electrical energy conversion step.

Bio-inspired manganese clusters have been shown to possess water oxidation activity when adsorbed on clays together with ruthenium photosensitizers, although with low turnover numbers.

Some ruthenium complexes are able to oxidize water under solar light irradiation. Although their photostability is still an issue, many can be reactivated by a simple adjustment of the conditions in which they work. Improvement of catalyst stability has been tried resorting to polyoxometalates, in particular ruthenium-based ones. Another way to achieve improved stability may be the use of robust clathrochelate ligands that stabilize high oxidation states of metal in catalytic intremediates.

Whereas a fully functional artificial system is usually intended when constructing a water splitting device, some mixed methods have been tried. One of these involve the use of a gold electrode to which photosystem II is linked; an electric current is detected upon illumination.

Hydrogen-Producing Artificial Systems

A H-cluster FeFe hydrogenase model compound covalently linked to a ruthenium photosensitizer. The ruthenium complex absorbs light and transduces its energy to the iron compound, which can then reduce protons to H_2.

The simplest photocatalytic hydrogen production unit consists of a hydrogen-evolving catalyst linked to a photosensitizer. In this dyad assembly, a so-called sacrificial donor for the photosensitizer is needed, that is, one that is externally supplied and replenished; the photosensitizer donates the necessary reducing equivalents to the hydrogen-evolving catalyst, which uses protons from a solution where it is immersed or dissolved in. Cobalt compounds such as cobaloximes are some of the best hydrogen catalysts, having been coupled to both metal-containing and metal-free photosensitizers. The first H-cluster models linked to photosensitizers (mostly ruthenium photosensitizers, but also porphyrin-derived ones) were prepared during the early 2000s. Both types of assembly are under development to improve their stability and increase their turnover numbers, both necessary for constructing a sturdy, long-lived solar fuel cell.

As with water oxidation catalysis, not only fully artificial systems have been idealized: hydrogenase enzymes themselves have been engineered for photoproduction of hydrogen, by coupling the enzyme to an artificial photosensitizer, such as $[Ru(bipy)_3]^{2+}$ or even photosystem I.

NADP$^+$/NADPH Coenzyme-Inspired Catalyst

In natural photosynthesis, the NADP$^+$ coenzyme is reducible to NADPH through binding of a proton and two electrons. This reduced form can then deliver the proton and electrons, potentially as a hydride, to reactions that culminate in the production of carbohydrates (the Calvin cycle). The coenzyme is recyclable in a natural photosynthetic cycle, but this process is yet to be artificially replicated.

A current goal is to obtain an NADPH-inspired catalyst capable of recreating the natural cyclic process. Utilizing light, hydride donors would be regenerated and produced where the molecules are continuously used in a closed cycle. Brookhaven chemists are now using a ruthenium-based complex to serve as the acting model. The complex is proven to perform correspondingly with NADP+/NADPH, behaving as the foundation for the proton and two electrons needed to convert acetone to isopropanol.

Currently, Brookhaven researchers are aiming to find ways for light to generate the hydride donors. The general idea is to use this process to produce fuels from carbon dioxide.

Photobiological Production of Fuels

Some photoautotrophic microorganisms can, under certain conditions, produce hydrogen. Nitrogen-fixing microorganisms, such as filamentous cyanobacteria, possess the enzyme nitrogenase, responsible for conversion of atmospheric N_2 into ammonia; molecular hydrogen is a byproduct of this reaction, and is many times not released by the microorganism, but rather taken up by a hydrogen-oxidizing (uptake) hydrogenase. One way of forcing these organisms to produce hydrogen is then to annihilate uptake hydrogenase activity. This has been done on a strain of Nostoc punctiforme: one of the structural genes of the NiFe uptake hydrogenase was inactivated by insertional mutagenesis, and the mutant strain showed hydrogen evolution under illumination.

Many of these photoautotrophs also have bidirectional hydrogenases, which can produce hydrogen under certain conditions. However, other energy-demanding metabolic pathways can compete with the necessary electrons for proton reduction, decreasing the efficiency of the overall process; also, these hydrogenases are very sensitive to oxygen.

Several carbon-based biofuels have also been produced using cyanobacteria, such as 1-butanol.

Synthetic biology techniques are predicted to be useful for this topic. Microbiological and enzymatic engineering have the potential of improving enzyme efficiency and robustness, as well as constructing new biofuel-producing metabolic pathways in photoautotrophs that previously lack them, or improving on the existing ones. Another topic being developed is the optimization of photobioreactors for commercial application.

Employed Research Techniques

Research in artificial photosynthesis is necessarily a multidisciplinary topic, requiring a multitude of different expertise. Some techniques employed in making and investigating catalysts and solar cells include:

- Organic and inorganic chemical synthesis.
- Electrochemistry methods, such as photoelectrochemistry, cyclic voltammetry, electrochemical impedance spectroscopy Dielectric spectroscopy, and bulk electrolysis.
- Spectroscopic methods:
  - fast techniques, such as time-resolved spectroscopy and ultrafast laser spectroscopy.
  - magnetic resonance spectroscopies, such as nuclear magnetic resonance, electron paramagnetic resonance.
  - X-ray spectroscopy methods, including x-ray absorption such as XANES and EXAFS, but also x-ray emission.
- Crystallography.
- Molecular biology, microbiology and synthetic biology methodologies.

Advantages, Disadvantages and Efficiency

Advantages of solar fuel production through artificial photosynthesis include:

- The solar energy can be immediately converted and stored. In photovoltaic cells, sunlight is converted into electricity and then converted again into chemical energy for storage, with some necessary loss of energy associated with the second conversion.

- The byproducts of these reactions are environmentally friendly. Artificially photosynthesized fuel would be a carbon-neutral source of energy, which could be used for transportation or homes.

Disadvantages include:

- Materials used for artificial photosynthesis often corrode in water, so they may be less stable than photovoltaics over long periods of time. Most hydrogen catalysts are very sensitive to oxygen, being inactivated or degraded in its presence; also, photodamage may occur over time.

- The cost is not (yet) advantageous enough to compete with fossil fuels as a commercially viable source of energy.

A concern usually addressed in catalyst design is efficiency, in particular how much of the incident light can be used in a system in practice. This is comparable with photosynthetic efficiency, where light-to-chemical-energy conversion is measured. Photosynthetic organisms are able to collect about 50% of incident solar radiation, however the theoretical limit of photosynthetic efficiency is 4.6 and 6.0% for C3 and C4 plants respectively. In reality, the efficiency of photosynthesis is much lower and is usually below 1%, with some exceptions such as sugarcane in tropical climate. In contrast, the highest reported efficiency for artificial photosynthesis lab prototypes is 22.4%. However, plants are efficient in using CO_2 at atmospheric concentrations, something that artificial catalysts still cannot perform.

Artificial Photosynthesis Applications

Fossil fuels are in short supply, and they're contributing to pollution and global warming. Coal, while abundant, is highly polluting both to human bodies and the environment. Wind turbines are hurting picturesque landscapes, corn requires huge tracts of farmland and current solar-cell technology is expensive and inefficient. Artificial photosynthesis could offer a new, possibly ideal way out of our energy predicament.

For one thing, it has benefits over photovoltaic cells, found in today's solar panels. The direct conversion of sunlight to electricity in photovoltaic cells makes solar power a weather- and time-dependent energy, which decreases its utility and increases its price. Artificial photosynthesis, on the other hand, could produce a storable fuel.

And unlike most methods of generating alternative energy, artificial photosynthesis has the potential to produce more than one type of fuel. The photosynthetic process could be tweaked so the reactions between light, CO_2 and H_2O ultimately produce liquid hydrogen. Liquid hydrogen can be used like gasoline in hydrogen-powered engines. It could also be funneled into a fuel-cell setup, which would effectively reverse the photosynthesis process, creating electricity by combining hydrogen and oxygen into water. Hydrogen fuel cells can generate electricity like the stuff we get from the grid, so we'd use it to run our air conditioning and water heaters.

One current problem with large-scale hydrogen energy is the question of how to efficiently – and cleanly – generate liquid hydrogen. Artificial photosynthesis might be a solution.

Methanol is another possible output. Instead of emitting pure hydrogen in the photosynthesis process, the photoelectrochemical cell could generate methanol fuel (CH_3OH). Methanol, or methyl alcohol, is typically derived from the methane in natural gas, and it's often added to commercial gasoline to make it burn more cleanly. Some cars can even run on methanol alone.

The ability to produce a clean fuel without generating any harmful by-products, like greenhouse gasses, makes artificial photosynthesis an ideal energy source for the environment. It wouldn't require mining, growing or drilling. And since neither water nor carbon dioxide is currently in short supply, it could also be a limitless source, potentially less expensive than other energy forms in the long run. In fact, this type of photoelectrochemical reaction could even remove large amounts of harmful CO_2 from the air in the process of producing fuel.

References

- Photosynthesis: newworldencyclopedia.org, Retrieved 14 July, 2019

- Factors-affecting-photosynthesis, photosynthesis-in-higher-plants: toppr.com, Retrieved 26 August, 2019

- Chlorophyll: newworldencyclopedia.org, Retrieved 22 April, 2019

- Modern-applications: historicalcasestudyphotosynth.weebly.com, Retrieved 05 January, 2019

- Artificial-photosynthesis, green-tech-energy-production: science.howstuffworks.com, Retrieved 12 July, 2019

- David Oakley Hall; K. K. Rao; Institute of Biology (1999). Photosynthesis. Cambridge University Press. ISBN 978-0-521-64497-6. Retrieved 3 November 2011

Permissions

Index

www.ingramcontent.com/pod-product-compliance
Lightning Source LLC
Chambersburg PA
CBHW080407190526
45161CB00003B/162